Mexico and the Survey of **Public Lands**

Mexico
and the Survey of
Public Lands
The Management of Modernization
1876–1911

Robert H. Holden

mm NORTHERN ILLINOIS UNIVERSITY PRESS
DeKalb 1994

© 1994 by Northern Illinois University Press

Published by the Northern Illinois University Press,

DeKalb, Illinois 60115

Design by Julia Fauci

Library of Congress Cataloging-in-Publication Data

Holden, Robert H.

  Mexico and the survey of public lands : the

management of modernization, 1876–1911 / Robert H. Holden.

    p.   cm.

  Includes bibliographical references and index.

  ISBN 0–87580–181–1

  1. Surveying—Public lands—Mexico—History.   I. Title.

TA622.H64   1994

333.1′6′0972—dc20   93–34379

                                                            CIP

*My daughter Sarah's presence every step of the way
made a hard job much easier.*

This book is for her.

Contents

Tables

Preface

SOURCES

The interpretations and conclusions presented in this volume are based primarily on documents that I read during a seven-month investigation in 1984–85 in the Archivo de Terrenos Nacionales of the Secretaría de Reforma Agraria in Mexico City. I also made use of records stored in the following Mexico City documentary repositories: Archivo General de la Nación, the Colección General Porfirio Díaz at the Universidad Ibero-americana, the Archivo General de Notarías del Distrito Federal, and the Archivo Histórico of the Secretaría de Relaciones Exteriores.

The Archivo de Terrenos Nacionales (ATN), however, provided the bulk of the data, and my work in that archive consumed by far the most time. I worked exclusively in the *archivo antiguo* section, whose records are largely limited to those produced before 1911, although I found correspondence written as recently as the 1970s in its files. Located in the basement corner of a multi story building occupied by the secretariat at Bolívar 145, the archivo antiguo of the ATN seems to have been generally off-limits to researchers. Virtually all of the nineteenth- and twentieth-century records of public land transfers were stored and updated in the archivo antiguo, a working repository attended by a full-time staff of five people. Less than three months after I finished working in the archive, the building was severely damaged in the 1985 earthquake, and some of the records may have been destroyed; many of those that survived the earthquake, I am told, were moved to a storage facility.

The ATN's archivo antiguo appeared to contain all of the federal executive branch documents generated by the surveys, which had been supervised by the Secretaría de Fomento. In addition to the survey documents, papers generated by other activities connected to the disposition of the public lands—such as the colonization program and records of land claims—were deposited in the ATN. Tens of thousands of files classified by state were organized according to *deslindes* (surveys), *colonias* (colonies), *baldíos* (vacant land), *permisos a extranjeros* (license for foreigners), *nacionales en compra* (public land sales), and so on.

For the purposes of this project, research was limited to files classified as deslindes and contratos originales. These files contained an extremely wide range of papers produced by the surveys, including documents written by Fomento officials, by the survey contractors, by landholders in the areas affected by the surveys, and by the judicial authorities who supervised the surveys. Copies of contracts, descriptions of land surveyed, official court records of litigation, and correspondence among Fomento, the surveyors, and landholders were the principal sources for this project, which to my knowledge was the first systematic investigation of the survey companies not limited to secondary sources.

Because of the sheer volume of documents in the ATN, I decided to limit my research to six states. This was nearly half the total number of states and territories (thirteen) in which survey companies received compensation. I picked Chihuahua, Durango, Sinaloa, Sonora, Chiapas, and Tabasco because they were among the top nine states in percentage of land awarded in compensation and because of the geographical balance that this selection furnished. Of the 21.1 million hectares given to all the companies in compensation from 1878 to 1908, those operating in the six states accounted for about half, or 10.5 million; see Table 1 and Table 9. The evidence on which the conclusions of this study are based, therefore, is limited to these six states. But together they account for almost exactly half of all the land given to survey companies in compensation during the Porfiriato. They therefore provide a reliable and balanced sample of survey company activity, in both quantitative and geographical terms.

Bias

At certain points in this book I will offer generalizations about the legality or fairness of the survey proceedings. A reader may reasonably object that even those records indicating that the proceedings were technically lawful were not necessarily truthful, and that even if they were truthful, they are not necessarily evidence that the procedure was fair to all property claimants. This challenge cannot be completely refuted, since a written record is only a written record, composed more often than not by authorities permanently vulnerable to charges of venality and self-interest. Unfortunately, neither the authors nor the supposed victims can be cross-examined. The bias against the illiterate, the voiceless, the weak, and the poor is one common to all written records. If no evidence what-

soever were found that poor and illiterate people ever took action to defend themselves against the survey companies, this bias would have
seriously undermined the findings of this study. Throughout the pages of
this work, however, scores of such actions are recorded, particularly in
chapter 4. Indirect evidence is also presented, such as references by others
to grievances. The bias, therefore, is not a crippling one, nor is it enough
to weaken the book's main arguments.

Like any historical archive the ATN's archivo antiguo is (or perhaps
was) incomplete in other respects. It could only contain data reduced to
writing. Further, an employee of the ministry had to properly file the
written document in the deslindes classification for it to be found by an
investigator whose main concern was the survey companies. Grievances
against the companies, action by the federal government, or responses by
the companies that were not written and communicated to the Secretaría
de Fomento, and then properly filed, were obviously not available. Some
communities and individuals who either lost land or feared losing it may
never have remonstrated, either in the presence of a local judge or to Fomento. Victims of usurpation could have failed to act because of ignorance or poverty; illiteracy could have prevented them from submitting
the necessary petitions, while lack of funds could have placed lawyers,
scribes, and even the time needed to travel to a court beyond their reach.
Moreover, others surely submitted written protests that were lost, destroyed, or otherwise unrecorded. Many thousands of documents were
reviewed for this study, but the findings are biased in the same way that
all studies based on written records are biased—in favor of the evidence
that exists in the records that happen to be preserved.[1]

How, then, does one justify the use of the official records of the Secretaría de Fomento to evaluate the fairness or accuracy of the surveys that
the agency itself was charged with supervising? First, the documents in
any given survey file originated in a mixture of sources: federal district
courts, local courts, survey companies and their employees, property
claimants, Fomento bureaucrats, and other government authorities. Since
their provenance varied widely, the records reflect more than just the
viewpoint of the companies themselves, or the government, or the property holders. This is the principal strength of the Fomento archives.

Moreover, the documents report enough evidence of conflict among
the various interests represented by the surveys to indicate that they do
reflect a wide assortment of interests. Survey companies, judges, lawyers,
landholders, Fomento officials, state governors, and *jefes políticos* (substate
political authorities) have their say and are by no means in accord with
one another, as will become apparent. Abundant critical commentary directed against the survey companies from a variety of elite quarters will

be reported. And as aforementioned, the poor and illiterate frequently speak through petitions that were preserved in Fomento's files to Fomento and to judges. They will be cited often in this book.

THE USE OF SPANISH

As a contribution to the public land history of Mexico that happens to be written in English, this book undoubtedly reflects some of the ambiguities that accompany any attempt to deploy concepts in two languages. In an effort to minimize the ambiguities, I have chosen to consistently use certain terms in Spanish throughout the text of this book:

> *baldío:* Untitled land, by definition the property of the nation, or public land.
>
> *cantón:* A substate jurisdiction, equivalent to a *distrito* (district).
>
> *compañía deslindadora:* A company hired by the state to survey land.
>
> *composición:* A division of public and private properties that surveyors were often empowered to negotiate with landholders, who secured clear titles as a result.
>
> *demasía:* The area by which a polygon described in a title exceeded the quantity specified in the title. For example, a title might concede one thousand hectares to an individual, but upon surveying the property within the limits described in the document, an engineer might discover that the area actually contained, say, fifteen hundred hectares. The demasía, in this case, amounted to five hundred hectares. Such land was considered more or less equivalent to baldío; it could be subject to a claim or composición.[2]
>
> *ejido:* Generally, land held in common by rural agricultural producers.
>
> *fundo legal:* Land reserved for public purposes by a municipality.
>
> *Secretaría de Fomento:* The cabinet ministry in charge of public works, including the disposition of the public lands.
>
> *terreno nacional:* Formerly baldío land that had been surveyed and was therefore properly titled to the federal government. It consisted of the government's two-thirds share of surveyed baldío, and was sold at a negotiated price by the government.

THE VALUE OF LAND

Throughout most of this book, reference will be made to land as if it were a homogeneous commodity. Survey companies will be compared on the basis of the quantity of land they received in compensation, conflicts over land between the companies and landholders will be analyzed, and so on. Of course all land is not the same; a 100-hectare patch of river-fed real estate in Sinaloa could have been worth many times more than 10,000 hectares of Chihuahua desert. Unfortunately, the quality of the land in question in each case was something that could rarely be known. I have tried to avoid comparing pieces of property merely on the basis of size. An analysis that takes into account land quality and potential will have to await the discovery of far more detailed data than I was able to find. I offer chapter 5, in which I treat the question of land prices (and therefore perceived value) at some length, as partial reparation for any damage done.

MEASURES OF LAND[3]

1 *sitio de ganado mayor* = 1,756 hectares
1 *sitio de ganado menor* = 780 hectares
1 hectare = 2.47 acres

MONEY

The $ sign alone refers to Mexican pesos; U.S. dollars are denoted as US$. Where the monetary unit is ambiguous the sign ?$ is used. The Mexican peso traded one-for-one with the U.S. dollar until 1872, when the peso began its gradual descent—to US$0.91 in 1878, US$0.76 in 1888, and US$0.45 in 1898.[4]

ARCHIVAL CITATIONS

ATN: Archivo de Terrenos Nacionales Documents were filed according to two different classification systems. Under both systems, the name of the state comes first; I used the following state abbreviations: CHIAP = Chiapas; CHIH = Chihuahua; DUR = Durango; SIN = Sinaloa; SON = Sonora; TAB = Tabasco. Under the decimal system of classification (such as, for example, CHIAP 1.71/1) the documents were often paginated within the file cited, and when they were I gave the page numbers only. Wherever documents were not paginated, such as in the nondecimal classification (for example, CHIH 75669), I have cited the document by name and added the date if there was a date. The word *memo* refers to an interoffice memorandum produced within the Secretaría de Fomento. *Court file* refers to the survey documents generated by the local judicial authority that oversaw the survey, usually a federal district court.

NOT: Archivo Histórico of the Archivo General de Notarías del District Federal

CPD: Colección General Porfirio Díaz, located in the library of the Universidad Iberoamericana, Mexico City

AGN: Archivo General de la Nación

SRE: Archivo Histórico de la Secretaría de Relaciones Exteriores

ACKNOWLEDGMENTS

This book first took shape as a series of seminar papers produced under the direction of two great teachers, John H. Coatsworth and Friedrich Katz, who later supervised the dissertation on which this book is based. I am deeply indebted to them for their encouragement and advice. Enrique Florescano provided indispensable assistance in gaining access to the Archivo de Terrenos Nacionales. Brígida von Mentz, Donald F. Stevens, and Hector Lindo Fuentes were among the fellow scholars of Mexico and Latin America whose thoughtful criticism and suggestions were always a source of stimulation. I hope that no one will be disappointed by the results; I alone take responsibility for any shortcomings.

It is a pleasure to acknowledge with gratitude the generous financial assistance of the University of Chicago; the Center for Latin American Studies of the same university; the U.S. Department of Education, which supplied a Fulbright-Hays Doctoral Dissertation Research Abroad Award, as well as a National Resources Fellowship (Title VI); and the Inter-American Foundation, for a dissertation research fellowship. Duke University Press generously permitted me to use portions of an article I wrote for the *Hispanic American Historical Review*.

Encouragement and support of a less tangible nature during the research and writing phases of this project were lovingly and abundantly furnished by Nacho and Elena González, Zack Schiller and Gail Long, George Hrbek and Stephanie Morrison, Jon and Donna Halvorsen, Rina Villars Rosales, Warren Stork, Pauline Holden Stork, and Lynn Holden.

Mexico and the Survey of Public Lands

This monkey said to his friend the cat,
"I'm sure you know what I'm looking at—
Those lovely chestnuts there in the fire.
But I would be a miserable liar
If I declared that I was a better
Delicious-red-hot-chestnut-getter
Than you. For no one ranks any higher
Than you when it comes to the handling of fire.
If you'll pluck them out and quickly slide them,
I'll sit over here and fairly divide them."
So the flattered cat stretched out his paw
And plucked at an ember with just one claw,
Then quickly drew his paw back in pain.
But bravely he tried again and again,
And though at last he clearly had learned
How to get chestnuts, he still got burned.

—"The Monkey and the Cat,"
Fables from Aesop by Ennis Rees

The state government of Morelos would be ill-advised to carry out surveys of public land on its own account. The administration would run "all the risks consequent to its boldness because it is going to do by itself and under its own social responsibility what can be done by means of the Survey Companies: that is, it would have to pull out with its own hand the chestnut that it could pull out with the cat's paw."

—President Porfirio Díaz to
Jesús H. Preciado, governor of Morelos,
February 17, 1890.

1 Land and the State in Prerevolutionary Mexico

 Between Mexico's tempestuous first half-century (1821–76) and the bloody insurrections of 1910–17 reposes the age of Porfirio Díaz, the general and president whose first name, like Napoleon Bonaparte's, has come to designate the era of his rule: the Porfiriato.

During those three and one-half decades, Mexico's political economy developed features immune even to the revolution to which it would give birth in 1910: the consolidation and centralization of governing authority in Mexico City, the monopolization of political power by a single party, the utilization of the state as both engine and pilot of economic growth, and the growth of an export-oriented economy driven largely by foreign lenders and investors whose needs the new state sought to gratify. To establish the context both for these secular trends and for the public land surveys that are the subject of this book, a brief review of the main socioeconomic characteristics of the Porfiriato is in order.

Like all of Latin America and much of the rest of today's Third World, Mexico during the last quarter of the nineteenth century was drawn into that epoch-forming global drama that E. J. Hobsbawn called the "age of capital." Capitalism would engulf the world in the twentieth century, seeking to overcome the natural limits and the manmade barriers to the movement of labor, goods, and capital.

By the time Porfirio Díaz took office in 1876, Mexico had already begun to respond. But more than a half-century of violently unstable government, foreign invasions, rural insurgencies, and the loss of more than half its territory to the United States had left Mexico both weak and vulnerable to its colossal northern neighbor. To benefit from the valuable natural resources that its leading citizens were convinced lay within the

country's now-shrunken borders, Mexico would need to perform a difficult feat: maintain a strong, stable central government capable of defending national sovereignty while encouraging foreign economic investment
and protecting domestic entrepreneurs from foreign competition. Its
government would need to win the confidence not only of the militarized
factions competing for political power but also the overseas governments,
traders, and investors whose cooperation was crucial.

In large part, this task was achieved between 1876 and 1910, although
not without costs that would come due in the violence of 1910–17. During the presidencies of Díaz (1876–80; 1884–1911) and Manuel González (1880–84), Mexico finally gained political stability and fashioned a
state committed to systematic intervention on behalf of economic
growth. Responding largely to the prodding of the state and to a variety
of trends overseas, the economy shook out of its decades-long stagnation
and expanded at a spectacular rate.[1] Like so many epoch-making trans-
formations in human history, however, those of the Porfiriato emerged as
the result of civil war.

THE STATE AND ECONOMIC GROWTH IN PORFIRIAN MEXICO

The decisive victory of the forces of liberalism under the leadership of
Benito Juárez in 1867 buried forever the hopes of those conservatives
who sought a restored monarchy, a Catholic church with its ancient
privileges intact, and a rigidly controlled economy operating for the exclusive benefit of the landed oligarchy. But no liberal government could
survive for long on principles alone, and so there arose a tradition of political compromise with elements of the defeated conservatives as well as
with regional strongmen. Juan Felipe Leal has called it the "liberal oligarchic state," a label that neatly encapsulates the paradox of the state
that would emerge after 1867. Formally democratic, with a division of
powers and devoted to the principles of free trade and open markets, in
its actual behavior the government continued to serve the needs of the
most important regional oligarchs just as it simultaneously sought an
exit to a liberal capitalist society. The oligarchs' persistent attachment to
noncapitalist forms and practices such as peonage, as well as the despotic
rule of regional strongmen (*caciques* and *caudillos*), required patience and
creativity on the part of the progressive, capitalist-oriented governing
authorities.[2]

No one proved more successful at managing this contradictory state of
affairs than Porfirio Díaz, the liberal general and hero of the Battle of
Puebla. Rejecting the policy of exclusion that had paralyzed government
since the liberal victory, by the late 1870s Díaz had become the great con-

ciliator, opening up the Chamber of Deputies to opposition leaders and even providing cabinet posts to individuals who had collaborated with the foreign usurper, the emperor Maximilian. Some liberals who had opposed Díaz's Plan de Tuxtépec entered the government. Besides a policy of conciliation, Díaz also mastered the art of balancing local and regional interests without directly confronting them. Allowing the chamber to immobilize itself in constant debate while making a show of his own practical skills at the balancing game, Díaz concentrated power in the executive branch. Eventually, he would assert his authority over the multiplicity of regional and local powers by turning them against one another, and then providing sinecures in compensation.[3] Juan Felipe Leal explains how the dictatorship's well-known battle cry reflected its political behavior:

> The phrase 'a little politics and lots of administration' was something more than a slogan. It was the sign of a political reality. Within this framework, Porfirio Díaz appeared, ideologically, as the supreme arbiter of the country, and in fact followed a policy of gauging the internal forces that were in conflict and announcing his decision, which was followed even by those who were damaged by it. All of this, within narrow oligarchic limits.[4]

The "liberal oligarchic state" sought above all to create the conditions necessary for the operation of a capitalist society and to modify or better yet eliminate practices that interfered with the new order. In the absence of a national bourgeoisie, the state was compelled initially to seek additional support for its program of promoting a new capitalist order by creating political institutions that were formally liberal and democratic, such as a democratically elected congress, and by permitting the operation of a fairly free press. But such conditions were not conducive to the stability and control required by what was, in effect, a revolutionary state, or a state with a revolutionary program. Thus political liberties were exercised in a meaningful way only briefly, until about 1880; from then onward, they were gradually restricted as the state became more and more authoritarian, a trend that coincided with the expansion of capitalism. In the absence of promising alternative channels such as a congress, political parties, or elections, the hegemony of any given sector of the ruling class could only be attained by means of a direct relationship with the presidency, where power was concentrated. Such hegemony was limited only by the constraints imposed by the presidential strategy of maintaining a balanced system of regional alliances, for this system made it possible for Díaz to keep power.[5]

The immense authority that would be concentrated in the national state allowed it to become the leading advocate of capitalist enterprise, with its participation in the economy advancing far beyond the limits prescribed by orthodox liberalism. Indeed, the Mexican state was in a commanding economic position from the very start of the Porfirian era. John H. Coatsworth estimates that government spending accounted for about 12 percent of Gross Domestic Product in 1877. On a per capita basis, government spending in constant pesos grew 43 percent from 1877 to 1910, rising from $5.29 to $7.55.[6] Its intervention in the economy expanded sharply in response to the requirements of capitalist expansion in the interior (in terms of support for mining, export agriculture, trade, banking, and industry), as well as those of foreign capitalists making direct investments in Mexico. The latter required modern port facilities and efficient and honest customs collection. To get the railroads built the government offered foreign capitalists exclusive rights to operate lines for ninety-nine years or more, and even subsidized construction costs, negotiating these and other guarantees individually with each contractor. Eventually, the Díaz regime carried out its most ambitious act of economic intervention by purchasing majority control of virtually all the rail lines and organizing a single national entity under its control.[7]

For Mexico, capitalist expansion in the age of capital would have been unthinkable without the generous participation of foreigners. The partnership would be mutually beneficial: Mexico needed capital, but capitalists also needed a strongly interventionist state capable of providing incentives and guarantees, security, and a certain amount of bureaucratic efficiency. The most powerful foreign influence in Mexican affairs was of course the United States, whose proximity and overwhelming superiority in wealth, technology, and population were both welcomed and feared. Mexico reacted by seeking to balance U.S. influence with more European capital, a strategy that backfired when some important U.S. interest groups eventually turned away from Díaz and sought allies from among the president's opponents. This need to manipulate competitive interests eventually acted as a severe constraint on the Mexican government; the country's emergence as a target of U.S.–European rivalry would contribute to the downfall of the Díaz government.[8]

The Díaz strategy paid rich dividends in economic growth, which has been estimated at 2.7 percent a year during the thirty-five-year Porfiriato—about double the population growth rate of 1.4 percent. The expansion was driven by foreign demand for Mexican exports, an abundance of foreign investment capital, and myriad technological advances, but facilitated above all by the adroit management of the dictatorship.[9]

THE PUBLIC LANDS AND THE SURVEY IMPERATIVE

In no sector of the economy (with the possible exception of the railroad industry) was the Mexican state's intervention more decisive, and its impact greater, than in the development of the public lands. Not only in Mexico but virtually throughout Latin America the land market erupted in response to soaring demand in the more developed countries for its agricultural and mineral products. Networks of newly constructed railroad lines further raised the value of land by sharply reducing transportation costs. To secure the maximum economic benefits for itself, for the emerging class of domestic entrepreneurs, and for foreign investors whose capital it sought to attract, the state intervened abruptly and massively in the land market. All over the continent, governments transferred to private hands enormous quantities of land hitherto held by the government itself, by the Catholic church, and by communities of indigenous peoples. The pace and timing of the interventions varied from country to country depending upon local conditions; the process was under way as early as 1850 in some places, and would play itself out by about 1920.

In the following pages I will review the public land history of independent Mexico up to 1882. I will then introduce the main theme of this book: the state's management of the survey of public lands by private companies that became both the agents and the beneficiaries of the single greatest movement of public lands in the nation's history.

The public land history of independent Mexico divides naturally at two points in time: in 1848, when the country lost most of its public lands to a foreign aggressor, and in 1882, the start of the period in which most of the remaining public lands were transferred to private individuals. As a result of the U.S.–Mexican War of 1846–48, Mexico ceded to the United States 55 percent of its territory under the Treaty of Guadalupe-Hidalgo (1848) and the Gadsden Treaty (1853). A total of 241,022,800 hectares (930,590 square miles) was severed and eventually became all or part of the states of Texas, Colorado, Kansas, New Mexico, Oklahoma, Arizona, California, Nevada, New Mexico, Utah, and Wyoming.[10]

Spurred by the realization that they had lost the northern territory because of their failure to exercise control over it, the liberals who took power in the Revolution of Ayutla of 1854–55 took the first tentative steps toward a coherent public land policy. Their most important act was to make the alienation of public lands strictly a federal responsibility; all alienations made by subnational authorities in any epoch were nullified, unless subsequently approved by the federal government.[11] This proclamation, however, did not address the most serious underlying obstacle to the alienation of Mexico's public lands: no one was quite sure where they were. In an early but futile attempt to develop a program of action, President Ignacio Comonfort's minister in charge of economic development, Manuel Siliceo, ordered all governors and *jefes políticos* to designate someone in every state and territory to survey all the *terrenos baldíos* (vacant or public lands).[12] Later, writing on the eve of the three-year War of the Reform, Siliceo took up the same question once again, reporting to the constitutional convention that little could be done with the public lands until the government determined their size and location. He then predicted the main impediment to such surveys, one that President Díaz would confront more than two decades later. A survey of private property would be required to identify the public lands, Siliceo said, "and it is certain that the most energetic resistance will be opposed to this, for reasons that cannot be hidden from the Sovereign Congress." Very few property owners do not believe that they own "that which may be found in the environs of their fincas."[13]

Surveys were necessary in part because of the unreliability of the titles held by most property owners. Siliceo traced the lack of correspondence between titles and boundaries to the freedom with which colonial-era conquistadores claimed land for themselves while making little effort to demarcate its limits. Crown-sponsored composiciones, or arrangements whereby land claims were legitimized by the payment of a fee, only confused property rights even more by legalizing the usurpation of public land. One result was frequent litigation. Titles examined by Siliceo's ministry turned out to concede "much less" land than was actually claimed under them—in some cases the discrepancy amounted to a factor of seven or eight.[14]

In April 1856, Siliceo signed a survey contract with a private company whose terms would become commonplace three decades later. Jecker, Torre y Compañía was authorized to survey the public lands of the territory of Tehuantepec; its sole compensation would be one-third of all the land it surveyed. Five months later, the company was authorized to undertake surveys on the same terms in Sonora and Baja California.[15] Again anticipating objections that would surface much later, Siliceo defended the terms of compensation on the grounds that the company would have to pay high salaries to its engineers, work in deserted areas that lacked all creature comforts, and face costs that would rise in proportion to the "resistance that private owners oppose to the verification of whatever land may be untitled [baldío], which necessarily makes them lose much more time than they had expected, obliging them at every step to file lawsuits requiring substantial expenses."[16] So much additional litigation was expected that the Ministerio de Justicia appointed special judges to accompany the surveyors. The judges were instructed to not let anyone occupying land, even without title, be deprived of it, pending appeal to the government "within a prudent period of time."[17]

The War of the Reform and then the French intervention afforded little opportunity for the implementation of those surveys or any other policy initiative, with one exception. With an eye toward raising revenue to fight the French, the war government of President Benito Juárez issued from San Luis Potosí on July 20, 1863, the *Ley Sobre Ocupación y Enajenación de Terrenos Baldíos* (Law on the Occupation and Alienation of Vacant Land). Establishing the means by which individuals could claim public land, this law controlled the claims process until its repeal in 1894. Individuals could submit claims for up to 2,500 hectares, surveying the land themselves and paying the price set biennially by the Ministerio (later Secretaría) de Fomento for each state. Anyone who resided on the land at least ten years would automatically be entitled to its ownership, at a reduced price.[18]

Following the restoration of the republic in 1867, the pace of alienation of the public lands under Juárez's law did not begin to increase sharply until after 1877. From 1867 to 1877, the highest level reached in any one year was 273,615 hectares (in 1876). Claims would not fall below that level again until 1892. Of the 45.7 million hectares of public land either sold outright by the government or given away in compensation to survey companies from 1867 to 1908, only 3.5 percent or 1.6 million hectares were transferred between 1867 and 1877.[19]

Not until after Porfirio Díaz took office would the rush for public land commence. Control of the public lands was the responsibility of the Secretaría de Fomento, which at the same time was pouring money into railroad construction, a telegraph system, colonization, and the modernization of agriculture and mining. More than one-third of the 1883–84 federal budget, for example, was assigned to Fomento. "With no established economic priorities, with no trained staff, the Department of Fomento plunged bravely—if blindly—into a large-scale program of government promotion of economic development."[20] The tremendous surge in public land transfers that took shape gradually in the late 1870s, peaked in the mid-1880s, and finally collapsed in the mid-1890s was administered by a government driven by two key objectives: rapid economic growth and the acquisition of more revenue to balance a chronically underfinanced budget.

The discovery and transfer of public lands would contribute to economic growth in two main ways. First, land owned by the government was by definition unproductive, except when it was being rented or temporarily exploited under a special permit, such as those issued for the cutting of timber. By privatizing it, the government provided investors with a resource that could be exploited, sold, and resold. Second, large-scale development of the country's natural resources was simply unfeasible without reliable titles and definitive property lines. Merely selling public land was not enough; it also had to be clearly differentiated from private property, and private properties from one another, for land to be a safe investment, free from the threat of constant litigation or worse.

The second objective, expanded state revenues to balance the budget, would be achieved by the sale of public land and by the anticipated economic growth. As the head of Fomento, Gen. Carlos Pacheco, noted in a memorandum to a subordinate in 1885, the government's goal was to "acquire the greatest number of lots possible."[21] A Fomento circular to all survey companies that same year urged them to finish their work "as soon as possible so that the Government can dispose of its share of the lots, mobilizing these public riches."[22] When a federal judge in Durango delayed sending the record of a survey to Fomento for its approval in 1887, a message from President Díaz to the Secretaría de Justicia urged that it be forwarded as soon as possible, "since both the Company and the Government suffer damages with the delay."[23]

National defense was also a factor in hastening the survey process. The development of land along the country's international borders, much of which was sparsely populated, was particularly important for security reasons. In 1890 the government reversed a decision to forbid survey company activities within one hundred kilometers of the Guatemalan border, reasoning that "it is a matter of the vitality of a frontier State of the Republic [i.e., Chiapas] that needs capital and intelligent workers in order to take advantage of the wealth of its extensive territory," especially in southern Chiapas, where "there are vast wildernesses which must be populated" to secure the border.[24]

Achieving these objectives called for two distinct efforts. The first consisted of the continuation of the public land claims procedure established by President Juárez in 1863, under which Mexico would dispose of 17.3 million hectares of public land between 1878 and 1908. (An additional 5.2 million hectares were sold between 1894 and 1908 under a separate procedure established for the sale of already-surveyed public land.) The second was the survey, under government auspices, of the public land. Surveys would identify which lands were truly public (i.e., untitled) and thus make it possible for the government to sell them. The surveys themselves would add value to the land, enabling the government to sell specified lots at prices higher than it could charge under the land claims program, which required the claimants to do their own surveys.

Until the early 1880s the government had done little to promote the exploitation of public land beyond encouraging the submission of claims as provided for in the 1863 law. Steps were taken to make it easier to submit claims, to end abuses by occupants of public land, and to ensure compliance with the legal requirements by claimants.[25] But when General Pacheco submitted a report to Congress on his department's activities from 1877 to 1882, he pointed out that the government's responsibility did not end there. Surveys were badly needed. Among the reasons he gave for the government's decision to attend more closely to the survey of property were the "spirit of enterprise" visible throughout Mexico and the anticipated increases in both the "public wealth" and treasury revenue.[26]

General Pacheco's report also sided with the growing number of voices raised against the chaotic state of property relations in the countryside. Just as in Siliceo's time, the government's principal concern was to locate the public lands to which it was entitled; the main obstacle continued to be the cloudy and disordered state of property relations in rural areas. The general observed that this disorganization constituted an obstacle to economic growth. Only a program of surveys could overcome it. So much confusion existed over the location of boundaries between private and public land that "it will not be an easy job to wipe it away in long years of arduous labor"; the task was long delayed in any case by the lack of sufficient government funds.[27] Another authority pointed out that property at the time was "a true chaos," with rural landholdings characterized by "great expanses unsurveyed in themselves, and including or confusing themselves with publicly owned land, obscure and unintelligible titles, hotbeds of litigation, lands of ancient and undivided communities, etc., etc."[28]

The vagueness of so many land titles caused endless arguments among property holders and then between the latter and the survey companies. When Jesús Valenzuela's company surveyed Chihuahua's Cantón Abasolo in 1884, it found Pedro Zuloaga's claim to 158,867 hectares for his Hacienda de Santa Catalina "absurdo." The surveyor described the problem in his report: "Without doubt it is one of the many cases in which the primordial titles cite as corners 'a hill' or 'a mountain range' or 'a stream' or 'a brook' or 'a savannah' and other similar natural landmarks or indefinite things, which can be found in any part of the surface of the world."[29] The governor of Coahuila in 1879, Gen. Hipólito Charles, urged Fomento to move quickly to survey the state's baldíos, both to clarify property lines and because of a recent surge in the creation of new towns and the demand for land in Coahuila from Texans eager to move there.[30]

Not only would the surveys distinguish the limits of private and public properties, but by turning government-authorized surveyors loose on the countryside the regime also hoped that it would have the effect of forcing property holders to perfect obscurely worded titles and to regularize their holdings.[31] Because the survey companies were instructed to identify all baldío or untitled land, and were free to designate one-third of what they found for their own compensation, the surveys would be a potent source of pressure on individuals who were squatting on public land. To protect their land, they would have to file a claim to it under the 1863 law. As President Díaz told the governor of Nuevo León in 1889, "there is no doubt that if the surveyors avail themselves of vast territorial extensions, it is because of the indolence of the citizens, who do not make haste to duly execute their claims." Small holders, the president said, should hasten to make use of the advantages provided by the law to those who are already occupying public land; in at least one case, he promised to hold off the work of a survey company until the holders of the land threatened by the company petitioned for title to it.[32]

The surveyors remained a useful lever in the continuing effort to clarify property relations. In 1896 a Fomento functionary noted that pressure from Rafael García Martínez's survey company was the best way to force holders of land in Villa Ocampo, Durango, to "perfect their titles," which action would also benefit the treasury. "It is almost certain that they will never try to legalize their titles if they do not feel obligated to do so, remaining indefinitely in possession of those lots so that the Government cannot receive anything for them." The opinion was seconded by Fomento's legal advisor, to whom "the most important thing" was the acquisition "for the Nation of that which legally and fairly belongs to it."[33]

This was the voice of the modern state, asserting its prerogatives with self-righteous certitude. Its objectives were understood and shared by the survey companies. The government, wrote survey contractor and lawyer Joaquín Casasús in 1888,

> has wanted not only to find out what land is in the possession of the Federal Government, but to avoid the unlawful occupation of rural property by landowners, regularizing and mapping what they presently possess. In fact, without the survey companies, landowners never would present themselves to the government in order to regularize their property; . . . Today when the threat exists that a Company is going to measure their land, when the frauds that have been committed have already been discovered, they make haste to present themselves to the Government. . . . The idea that the Ministry [of Fomento] was following is thus assured: the landowner regularizes his property, and ceases to be simply a squatter.[34]

Others considered this use of the survey companies to be counterproductive. One of the companies' severest critics was José Covarrubias, a high official of Fomento whose opinion, paradoxically, coincided with much of the anti-Díaz rhetoric of the period. After the Díaz government was overthrown in 1910, Fomento published a report that Covarrubias had submitted to Fomento secretary Leandro Fernández in 1903. Because perfecting their titles was so costly, landholders often did not bother to do so until "the danger was inevitable, when the speculator appeared" who had been authorized by the government. The government's strategy was unfortunate, Covarrubias wrote, because it provoked resistance in the countryside and preferred the perfection of titles to "the tranquillity of the possessors." In what might be characterized as an attitude of extreme pragmatism, Covarrubias argued that while clarity of land-ownership would be convenient, the objective should not have been allowed to interfere with agricultural production—landholders who could not afford to perfect their titles fell prey to "speculators," thus diminishing national income. Occupation and use, Covarrubias argued, should be "the best basis of property rights."[35] Though founded on the doubtful premise that "speculation" subverted economic growth, this view would form the basis of much of the rhetoric, if not the substance, of the postrevolutionary governments.

Besides the desire to encourage landholders to perfect their titles, the government also believed that the surveys would make the economy more productive. By releasing publicly held land for private exploitation, either by giving it away to the companies or by selling it for a few pesos per hectare, the backward and even primitive conditions in much of the country could be reversed. Some of Mexico's public land was often thought to be of so little value that even giving it away was better than letting it remain idle. In 1885 during a Chamber of Deputies debate provoked by opposition members who complained that the government was selling Chihuahua land too cheaply in comparison to the value of Texas land, a Díaz partisan described the natural wonders of Texas, then asked:

Can one say as much of the deserts and arid frontier lands of Chihuahua? Where do the many rivers flow that fertilize the land, and whose abundance serves as the motor of industry? Where does one find the social labor that the highways, the canals, railroads and centers of population represent? When was the substantial capital formed for the purchase of those lands that would determine the increase in their value? I do not need, Gentlemen, to describe the sad and devastating situation of the public lands of Chihuahua, which are completely devoid of the labor of man and of society, and which are subject to the terrible depredations of the savage.[36]

Not all of Mexico's public lands were as desolate as this description would imply, however. If they were, compensation of even one-third of the surveyed land would probably not have been enough to attract private surveyors. When Luis Huller's company surveyed part of the department of Soconusco in Chiapas in 1887, his surveyor reported that one could see "what splendors and attractions nature has granted these regions so that man, by means of intelligence and work, may transform these uncultivated forests and deserts into a future Eden that will be pleasant and happy."[37]

Merely by surveying the public lands, the Mexican government would also be moving a step closer toward that essential element of any rational system of land taxation—a cadastral survey. A report on Mexico prepared for the London-based Council of the Corporation of Foreign Bondholders, published in Mexico in 1887, pointed out that the country's internal *alcabalas* (customs duties) were the biggest obstacle to the development of trade. Among the substitutes proposed were a uniform federal tax on real property that would be fairly distributed among the states (the revenue from which would be devoted to abolishing the alcabala), and equalization of state real estate taxes. The report further indicated that "there is a serious obstacle to an equitable settlement of the tax on property, and that is the lack of a property register, and the enormous cost and the many years needed to prepare one. The transfer of property, from the hands of indigenous owners of fertile but uncultivated land, to foreign capitalists, is for them [the present owners] a source of 'apprehension.' "[38]

The revenue from additional public land sales would be welcomed by a government that by 1883–84 was passing through a "financial disaster" in the words of the Kozhevar report.[39] The increased value of surveyed land would enable the government to charge more and thus expand its revenues.[40]

THE PRIVATE SECTOR

And so it was that General Pacheco promised Congress that more attention would be given to the need to survey, and that for reasons of economy and others rather mysteriously unspecified, the government itself would not undertake the surveys. Private companies would be contracted to do the work:

Henceforth, the Government proposes to undertake the work of surveying in a more direct and effective manner, signing contracts with some businesses to carry out the surveys. From the point of view of economy of spending, as well as from other viewpoints of some importance that this Secretariat has had under consideration, and that will be made known at the appropriate time, this means of resolving the difficult problem of the survey of public and private property, may perhaps present fewer inconveniences than the great enterprise of surveying on the exclusive account of the Government about four-fifths of the territory of the Republic.[41]

Shortly afterward, in December 1883, the González administration obtained the passage of legislation authorizing it to hire private companies to do the surveys in return for one-third of the land they surveyed. The compensation formula was hardly a novelty by this time. As noted earlier, a precedent for undertaking surveys in this way had already been established in 1856 when Jecker, Torre y Compañía was hired to survey in Sonora, Baja California, and Tehuantepec, in return for one-third of the surveyed land. Moreover, eighteen months before Porfirio Díaz's *golpe* (coup) of November 1876, the government of President Sebastian Lerdo de Tejada had promulgated legislation (scarcely implemented) authorizing the use of survey companies and the one-third compensation formula.

One strong argument for hiring contractors to do the surveys was noted by General Pacheco—it would have been impossible for the government of Mexico in the early 1880s to pay for surveys when its budget deficits were chronic and growing. By hiring companies to do the work and paying them with land instead of cash, direct outlays would be minimal.

But there was another reason, and it was dictated by the parlous mission that the new state had set for itself of stimulating economic growth in a liberal capitalist framework while maintaining the support of the landed oligarchy. Verifying landholders' rights to the property they worked was a politically explosive undertaking. But hiring the services of private companies would be a sound way for the government to shield itself from the unpleasant consequences. This was undoubtedly the meaning of Pacheco's oblique reference to "other" reasons for hiring contractors. Public confirmation of the usefulness of this strategy was forthcoming at an early stage. When two Mexico City newspapers reported that property owners in Chihuahua, including the *latifundista* (large landowner) Luis Terrazas, were alarmed because surveyors were declaring their land baldío, Díaz ordered Fomento to inform all the companies that any indemnification for damages caused by them would be their own responsibility, and not the federal government's.[42]

But the best evidence was revealed in the president's private correspondence. When the governor of Morelos, Jesús H. Preciado, suggested to the president in 1890 a state law that would make the state responsible for surveys, an incredulous Díaz replied that not only did the idea conflict with the federal government's privileged jurisdiction over public lands, but that all the *"malevolencia y maldición"* ("malice and cursing") of the *pueblos* that were now putting up with the survey companies would be directed against the state. Díaz then spelled out, with a trace of exasperation, the political advantages of letting private companies open the tiger's cage: "that the survey companies carry that curse is not in the least important because when the situation becomes impossible, the Government deauthorizes them, obliges them to return [the land] and even punishes them if it is necessary to tranquilize the pueblos, something that cannot nor ought to be done with the Government of a State and even less when that Government is a friend." Díaz added that the proposed state law might be modified to make it more practical, but he urged the governor to bear in mind that the state government "runs all the risks consequent to its boldness because it is going to do by itself and under its own social responsibility what can be done by means of the Survey Companies: that is, *it would have to pull out with its own hand the chestnut that it could pull out with the cat's paw*" [emphasis mine].[43]

THE PUBLIC LAND TRANSFERS OF THE 1880S:
A DECADE OF TUMULT IN THE FRONTIER ZONES

General Pacheco's survey proposal, unlike that of his predecessor Siliceo, was not only immediately adopted by Congress but was actually implemented. It ignited the largest program of public land transfers in the nation's history; at no other time would any Mexican government privatize so much public land in so short a period. The contractors who were hired by the government to undertake the surveys not only acted as agents of the government but also were rewarded with one-third of the land that they chose to survey within the jurisdictions specified in their contracts. As a result of this compensation, the companies became the largest land developers in the country.

Their heyday was the decade of 1883–93, when the companies were given 18,484,964 hectares of public land in compensation for their surveys, or 87 percent of all the land that would be awarded in compensation during the entire Porfiriato. That decade's total compensation was equivalent to about one-tenth the total area of Mexico. If the companies indeed surveyed three times that number of hectares, they must have surveyed a staggering 55,454,892 hectares, or about 28 percent of the national territory—all in one decade! During the entire Porfiriato the survey companies received 21.2 million hectares of public lands in compensation (10.7 percent of the national territory), implying that 63.5 million hectares (32 percent of the national territory) had been surveyed under Pacheco's program.[44]

The scale of the survey company transactions is perhaps best grasped by comparing them to all the other methods by which the government disposed of public lands during the Porfiriato. Table 1 compares the annual privatization rates in total hectares. A total of 43.7 million hectares of public land was privatized between 1878 and 1908. The companies' share of that was 48 percent. Another 40 percent was transferred under the claims legislation of 1863 and 1894. The rest was newly surveyed land (i.e., the two-thirds left after the survey companies were compensated) that the government sold outright as terreno nacional. Of the 63.5 million hectares that were apparently surveyed based on the quantity of land given to the companies in compensation, by 1908 a total of 21.2 million hectares had been given to the companies and 5.2 million had been sold, leaving a balance of surveyed but still public land of 37.1 million hectares.

Transfers of public land under the claims laws as well as transfers to survey companies peaked in the 1880s (see Figure 1). Most of the activity was limited to just eight states and territories, as Table 2 indicates. Baja California, Campeche, Chiapas, Chihuahua, Sinaloa, Sonora, Tabasco, and Tepic saw between 19 and 47 percent of their land areas privatized, either as survey company compensation or under the claims laws. Together these regions accounted for 41 percent of the national territory, but they were the least populated in the country, lying in what was still frontier territory in the 1880s—the northern deserts, the northern Pacific coast, and the dense jungles of the South (see Figure 2). Table 3, which shows states and territories grouped by proportion of area alienated, indicates that, taken together, these eight jurisdictions had a population density as late as 1900 of just three persons per square kilometer. By comparison, population density in areas where no public land transfers occurred was eight times greater, or twenty-five persons per square kilometer. Even though the Mexican work force was still overwhelmingly engaged in agricultural activities, eight out of ten Mexicans lived in jurisdictions in which survey companies and public land claimants either did not operate at all, or did so at a very low level, as Table 3 shows.

Table 1   Annual Movement of Public Land, 1878–1908

Year	Baldío		Survey		Nacional	
	Titles	ha	Titles	ha	Titles	ha
1878	402	380,345	0	0	0	0
1879	325	420,895	0	0	0	0
1880	306	344,032	0	0	0	0
1881	432	461,948	0	0	0	0
1882	655	1,317,787	0	0	0	0
1883	532	3,087,102	—	1,665,312[a]	0	0
1884	513	5,635,901	—	1,665,312[a]	0	0
1885	531	797,478	—	1,665,312[a]	0	0
1886	332	557,882	—	1,665,312[a]	0	0
1887	262	551,619	—	1,665,312[a]	0	0
1888	337	612,404	—	1,665,312[a]	0	0
1889	342	415,534	—	1,665,312[a]	0	0
1890	297	368,858	—	1,665,312[a]	0	0
1891	373	375,116	—	1,665,312[a]	0	0
1892	219	242,726	—	1,665,312[a]	0	0
1893	161	277,435	—	1,831,848[b]	—	74,175[b]
1894	246	201,033	32	484,257	21	86,386
1895	67	45,060	29	243,576	19	59,265
1896	66	53,919	27	436,304	66	265,054
1897	108	84,170	7	562,093	75	230,932
1898	210	98,822	9	92,428	83	244,888
1899	140	179,232	1	4,263	214	276,986
1900	135	85,884	—	13,666	59	118,808
1901	112	98,113	19	50,534	121	225,712
1902	71	120,412	11	26,387	73	364,202
1903	85	45,253	7	42,705	132	576,473
1904	75	68,141	7	452,440	181	1,137,884
1905	125	125,093	22	146,556	132	533,707
1906	67	114,122	28	105,204	149	324,413
1907	122	103,162	23	21,366	106	390,546
1908	98	61,098	0	0	102	298,889
Total	7,746	17,330,576		21,166,743	1,533	5,208,320

Sources: Mexico, Secretaría de Fomento, *Memoria presentada al Congreso de la Union.* 1877–82, 1883–85, 1892–96, 1897–1900, 1905–7, 1907–8, 1908–9 (Mexico: Secretaría de Fomento); Mexico, Secretaría de Fomento, Dirección general de estadística, *Anuario estadístico de la república mexicana.* 1893, 1897, 1901, 1906 (Mexico: Secretaría de Fomento).

Notes: "Baldío" is land sold under the laws of 1863 and 1894 at government-set prices. "Survey" is land given to survey companies in compensation for surveying public lands. "Nacional" is surveyed public land, sold at a negotiated price.

[a]Annual average for 1883–92 in this category; source only gives total for this period.

[b]Fiscal year July 1, 1892–June 30, 1893.

Figure 1
Terreños Baldios and Nacionales Transfers, 1878–1908
(in hecters)

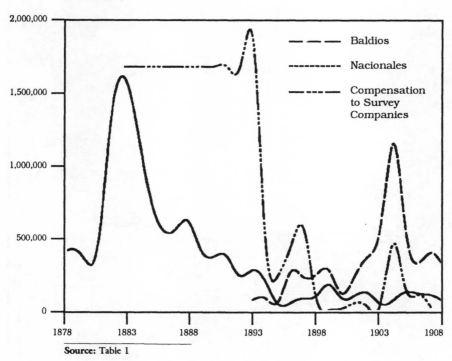

Source: Table 1

More public land was privatized in the form of survey company compensation than by any other method, as Table 1 indicates, and by the end of the Porfiriato about four times as much newly surveyed land had been given to the companies as had been sold outright by the government. About one-tenth of the national territory thus fell quite suddenly into the hands of probably not more than fifty private enterprises, primarily in the span of a single decade.[45] This estimate, which was derived from easily accessible data published by the governments of Díaz and González, inspires some provocative questions. Who were the survey companies? Where did they operate? What became of the land they surveyed? What happened to the land they acquired as compensation? Did the compensation method lead to the widespread theft of privately owned land that was coveted by the survey contractors? What was the economic impact of the surveys? How were they viewed by landholders

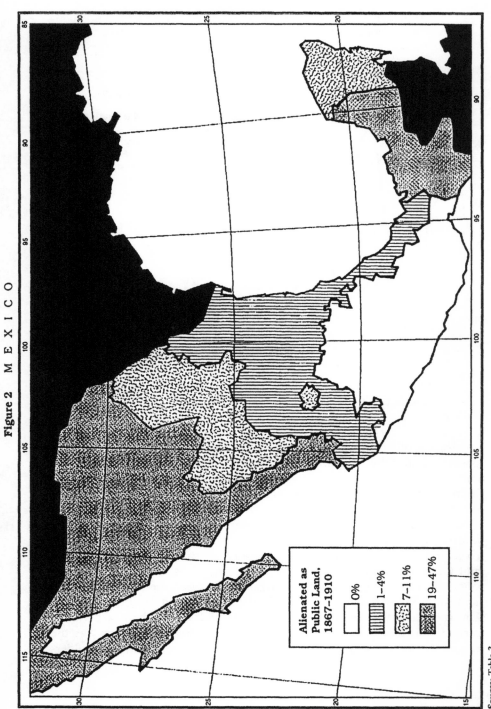

Figure 2 M E X I C O

Alienated as
Public Land,
1867–1910

0%

1–4%

7–11%

19–47%

Source: Table 3
Note: Boundaries reflect statewide averages only.

Table 2 Mexican Public Land Transfers, 1867–1909, Ranked by State

Sold (000 ha)		Survey (000 ha)		Total (000 ha)		Sold (%)		Survey (%)		Total (%)	
Sonora	2,598	Chihuahua	5,054	Chihuahua	7,550	Chiapas	35	B.C.	26	Tabasco	47
Chihuahua	2,496	B.C.	3,873	B.C.	5,105	Tabasco	31	Chihuahua	22	Chiapas	43
Chiapas	2,464	Sonora	1,515	Sonora	4,113	Campeche	20	Tepic	17	B.C.	34
Sinaloa	1,233	Chiapas	594	Chiapas	3,058	Sinaloa	17	Tabasco	16	Chihuahua	32
B.C.	1,232	Tepic	491	Sinaloa	1,390	Sonora	13	Chiapas	8	Sonora	21
Coahuila	1,126	Tabasco	430	Coahuila	1,299	Chihuahua	11	Sonora	8	Campeche	20
Campeche	963	Durango	411	Tabasco	1,269	B.C.	8	Durango	4	Sinaloa	20
Tabasco	839	Veracruz	242	Durango	1,209	Durango	7	Veracruz	3	Tepic	19
Durango	798	Coahuila	173	Campeche	963	Aguasc.	7	Sinaloa	2	Durango	11
Yucatán	574	Yucatán	166	Yucatán	739	Coahuila	7	Yucatán	2	Yucatán	8
Tamaulipas	287	Sinaloa	157	Tepic	546	Yucatán	6	S.L.P.	1	Coahuila	8
N.L.	219	S.L.P.	79	Tamaulipas	287	Tamaulipas	4	Coahuila	1	Aguasc.	7
Jalisco	89	Zacatecas	55	Veracruz	264	N.L	3	Zacatecas	0.9	Veracruz	4
Tepic	55	Others	0	N.L.	219	Tepic	2	Others	0	Tamaulipas	4
Aguasc.	55	—		S.L.P.	113	Jalisco	1	—		N.L.	3
S.L.P.	34	—		Jalisco	89	S.L.P.	0.6	—		S.L.P.	2
Veracruz	22	—		Zacatecas	55	Veracruz	0.3	—		Jalisco	1
Others	0	—		Aguasc.	55	Others	0	—		Zacatecas	0.9
—		—		Others		—		—		Others	
Total	15,083		13,240ᵃ		28,323						

Sources: Mexico, Secretaría de Fomento, *Memoria presentada al Congreso de la Union.* 1907–08, 1908–09. Mexico, Secretaría de Fomento, Dirección general de estadística, *Anuario estadístico de la república mexicana,* 1893, 1897, 1906; Mexico, Secretaría de Fomento, *Boletín semestral de la estadística de la república mexicana,* no. 3 (1889): 209–11.

Notes: "Sold" is total public lands sold as terrenos baldíos and terrenos nacionales, 1867–97 and 1901–9, in thousands of hectares. "Survey" is total public lands given to survey companies in compensation for surveying, 1881–89, 1894–97, and 1901–9, in thousands of hectares. "Total" is sum of first two columns, in thousands of hectares. "%Sold" is column 1 as percent of total area of state or territory. "%Survey" is column 2 as percent of total area of state or territory. "%Total" is sum of columns 4 and 5. Yucatán includes Quintana Roo; Distrito Federal is excluded. "Aguasc." is short for Aguascalientes, "B.C." is short for Baja California, "N.L." is short for Nuevo León, and "S.L.P." is short for San Luis Potosí.

ᵃTotal does not include an additional 2 million hectares distributed to individual contractors in the 1880s in parts of Coahuila, Chihuahua, Durango, Sonora, Tamaulipas, and Nuevo León that cannot be attributed to individual states.

in the countryside? What was the nature of the companies' relationship with the government?

Notwithstanding their extraordinary nature, the surveys have for decades been viewed by historians with a mixture of indifference and partisan contempt. The historiography regards the surveys as a shameful giveaway of the national patrimony to foreign capitalists, speculators, and

Table 3   States and Territories: Area Alienated and Population

	Group I (0%)	Group II (1–4%)	Group III (7–11%)	Group IV (19–47%)
	Colima	Jalisco	Aguascalientes	Sinaloa
	Guanajuato	San Luis Potosí	Yucatán	Tepic
	Guerrero	Zacatecas	Coahuila	Campeche
	Hidalgo	Nuevo León	Durango	Sonora
	México	Veracruz		Chihuahua
	Michoacán	Tamaulipas		Baja California
	Morelos			Chiapas
	Oaxaca			Tabasco
	Tlaxcala			
	Querétaro			
	Puebla			
Pop. share (1900)	53%	27%	8%	12%
Pop. density (per sq. km in 1900)	25	9	5	3
Share of area	18%	22%	19%	41%

Source: Table 2 and Mexico, Secretaría de Economía, *Estadísticas sociales del Porfiriato, 1877–1910* (Mexico, 1956), p. 68.

Note: Proportion of total area of state or territory alienated as baldío or terreno nacional, 1867–97 and 1901–9, and as compensation to survey companies, 1881–89, 1894–97, and 1901–9. Distrito Federal is excluded. Yucatán includes Quintana Roo.

big landowners. The process was said to have resulted in even greater concentration of landownership, in the theft of much Indian community land, the expropriation of small holdings, and therefore an unjust and "uneconomic" distribution of landed wealth. Because of its impact on the land tenure situation, the Díaz government's disposal of the public land is considered by many to be one of the principal causes of the Mexican Revolution of 1910.[46]

None of these assertions has ever been systematically tested. However, they have served the interests of the founders of the modern Mexican state of the "institutionalized revolution," a state whose legitimacy (like that of so many revolutionary states) derived in good part from its ability to disparage its predecessor. Enrique Florescano observed that the revolution of 1910 gave birth to a school of historiography that substituted "reasoned explanation with political argument, contributing to the creation of a black legend of the countryside and of the Porfirian hacienda that did not favor the study of their economic bases."[47] Francois Chevalier, a prominent historian of Mexico and a student of the land question, likewise observed that "this gigantic appropriation [by the survey companies] of *baldíos* or waste land has not been studied in a precise and systematic way. Its social consequences are still hardly known. Such research represents an urgent task for historians, especially in clarifying the origins of the Mexican Revolution."[48]

This book addresses the challenges posed by Florescano and Chevalier. I chose to focus on the survey companies because of their conspicuously pivotal role in the public land history of Mexico and because of the scarcity of basic information about their activities, despite their notoriety. Moreover, I believed that a study of this rather small group of private entrepreneurs would reveal something of the character of the state that hired them to undertake a task that was both politically risky and probably beyond the state's means financially.

I will argue that the survey companies had little if anything to do with the Mexican Revolution. The companies tended to respect titled property as well as land that was untitled but productively occupied, though they did not do so consistently. Evidence suggests that the government regulated the companies' activities quite closely (though again, not consistently), and defended landholders and disciplined survey companies in many cases of alleged abuse, though not in every case. Attempts by the postrevolutionary state to recover land that was allegedly stolen by the survey companies were little more than cynical devices to consolidate the regime's power. What all of this points to is a prerevolutionary state apparatus far more flexible and autonomous (at least through 1900), and more adept at balancing competing interests, than a reader of the historical literature might be led to think.

A facet of the Porfirian state that has often been asserted but rarely demonstrated is the extraordinary importance of the state's role in shaping Mexico's economic development in the three to four decades preceding the revolution. The state's commitment to following the path of the more developed capitalist nations was demonstrated throughout the period, as the governments of Manuel González and Porfirio Díaz acted boldly to improve communications and transportation, to open the public lands to development, to institute the legal norms and procedures that would encourage capitalists (especially foreigners) to invest in Mexico, and to keep the peace. In the disposition of the public lands, the state's role was clearly defined from the outset. The twin imperatives of raising revenue and stimulating economic growth required action to identify the public land and to turn much of it over to private ownership. The process, it was hoped, would also force individuals who held public land illegally to perfect their title to it. Unleashing private survey companies armed with a federal authorization was seen as the most rapid and economical way to identify the land and privatize it. When measured by the quantities of land surveyed and transferred to private hands, this policy turned out to be stunningly successful.

The next chapter will establish the legal and regulatory framework within which the companies were expected to operate, assess the effectiveness of that framework, and consider survey methods and cost-control incentives.

2 Fostering Development

The Mexican State and the Survey Business

The handful of partnerships and joint-stock companies that surveyed one-third of Mexico and received in compensation one-tenth of its area remain, more than a century later, largely unknown. Unfortunately, their obscurity is characteristic of the business class as a whole in prerevolutionary Mexico. The burst of literature on nineteenth-century regional themes in the past decade has yielded valuable information about certain prominent families and foreign entrepreneurs here and there. We know how some isolated sectors of the economy behaved in some years. But few historians have even begun to grapple with the question of entrepreneurship, let alone risk generalizing about the identity, character, motives, values, or political links of Mexico's entrepreneurial class before 1911. David W. Walker in his study of the Martínez del Río family led the way by proposing a theory of family, business, and politics in early modern Mexico. His conclusions, limited to the first half-century of independence, invite corroborative research. Roderic A. Camp's ambitious study of entrepreneurs and politics after the revolution stands at the opposite end of the time spectrum. Thus the Porfiriato, famous for producing what was probably the most concentrated burst of private sector activity in Mexico's history, still awaits the historian who can pull together the myriad strands of regional and sectoral trends into a coherent interpretation of entrepreneurship during the transition to dependent capitalism.

Walker and Camp attended primarily to the relationship between the business class and the state, showing how both before and after the Porfiriato, the public sector and the private sector fastened feverishly and abusively on each other.

In early modern Mexico, according to Walker, the state was the

submissive accomplice of the private sector, which "ruthlessly manipulated the state for private gain, to the detriment of class interests, economic growth, and political stability." The strong colonial state had
blasted every opportunity for individual initiative out of the economy
with its constant meddling, thoroughly politicizing the private sphere.
After independence the pattern persisted. But the state, now hopelessly
enfeebled, had become the victim of a group of entrepreneurs who fought
one another for the privilege of looting the state's resources, mainly by
speculating in its debt. This pitiful struggle was as economically unproductive as it was politically destabilizing, the two most prominent features of pre-Porfirian Mexico. Businessmen had not yet learned how to
act as a class on behalf of class interests, Walker concludes. The war-borne
liberal reforms corrected some of the structural weaknesses that had encouraged this state of affairs, and contributed to the opening of a new era
of political stability and economic growth.[1]

If before the Porfiriato a rapacious private sector had made the state its
mere accessory, after the revolution of 1910 the private sector had become the state's compliant ward, according to Camp. The triumphant
revolutionaries were far from being anticapitalist. But to stay in office
they needed to "polish their populist credentials" by keeping businessmen
out of government and by denying them affiliation with the official political party until the 1930s. By the 1980s the private sector had achieved
a measure of autonomy such that the relationship could be called "sym-
biotic." The private sector even acquired enough self-confidence to assert
that the historical strategy of state intervention should yield to market
forces, Camp pointed out.[2]

What, then, of the era of Porfirio Díaz? From the vantage point of the
twentieth century Camp gingerly probes for clues, speculating that Díaz
"gave too much power to Mexico's emerging capitalists" and thus excluded
the middle class, a fatal political mistake. The conceptual confusion between "emerging capitalists" and "middle class" deprives this hypothesis
of much of its value; one normally thinks of these categories as being
more or less synonymous. Camp cunningly resorts to the slippery rhetoric of the populist interpretation of the Mexican Revolution: "Emerging
capitalists" somehow transmogrified themselves into "the economic oligarchy." Whatever they were, they were very influential before 1910, he
concludes. Although they spoke in regional and personalistic voices
rather than national and institutional voices, "Mexico's capitalists were
more strongly represented in the Díaz government than in any period
after 1920."[3] That is, capitalists were more directly involved in the government of Mexico than at any time before (when they were incapable of
governing) or since (when they were were forbidden to govern).

Camp's own data bank, the Mexican Political Biography Project, suggests a less categorical judgment, however. While 29 percent of Díaz-era "political leaders" were the children of "ordinary landowners and businessmen," only 6 percent had parents who were "prominent" members of this class, a proportion that would decline to near invisibility between 1914 and 1935. In any case, Camp concludes, the participation of "prominent" families in the government during the Díaz era was insignificant. On the other hand, one out of five "leading politicians" during the Porfiriato was "an important businessman," and 17 percent of the era's leading politicians turned to business after leaving office. Thus while the level of exchange between the public and private sectors was higher during the Porfiriato than afterward, it was still not as high, say, as in the United States.[4]

A useful model of state–private sector relations in the Porfiriato would, in my view, emphasize first the regime's greatly revived capacity for *managing* the country's transformation to dependent capitalism and second its success in *rewarding entrepreneurs*, thus enlarging and enriching the class of investors at all levels of economic enterprise, from railroad and sugar mill magnates to small farmers.[5] François-Xavier Guerra captured the essence of this twin assault on the chaotic stagnation of the pre-Reform period:

> Once order was achieved, progress became the regime's watchword. It was as well to follow the example of the most advanced countries of the era. To get there, the liberal elite strengthened the instrument of its power, the State. . . . Porfirista policy was closer, in fact, to an 'enlightened despotism.' The expansion of the State appears in all spheres, manifested in the increase in public spending and the debt, in the progressive passage of economic legislation, in the growing control of education. This posture went hand in hand with a relative absence of social policy. The intervention of the State was destined to create modern economic actors. Once created, it was up to them to regulate their relations without outside interference. Out of this logic of the State came both its growing intervention over society and its social abstentionism. In the shadow of the State grew the social groups that were its servants: functionaries, teachers, judges and lawyers needed for disentailment, etc. It was among these groups that the men in power came to recruit their clients.[6]

In this chapter we will see how one well-defined sector of the entrepreneurial class—the firms that received land survey contracts in Chiapas, Chihuahua, Durango, Sinaloa, Sonora, and Tabasco—drew widely from among members of the *científico* elite (reformers associated with the Diaz regime), cash-rich American fortune hunters, and local businessmen and merchants. Profiles of twenty of these enterprises have been collected in Appendix A. Their relations with the state will be explored in some depth both here and in subsequent chapters.

Before attempting to analyze those relations, however, the legal and regulatory framework that more or less controlled the survey process must first be clarified. Second, an analysis of the distribution of survey contracts and compensation will give some sense of the competitive conditions under which the companies worked and an idea of their relative strength and success. Two interrelated issues will then be introduced and reviewed: survey techniques and the economics of the survey business. Finally, some generalizations (drawn from the profiles in Appendix A) will be offered about the identity of the companies themselves: Who were the men (and in at least one case, the women) behind the companies? How many were foreigners? What were their ties to the governing elite? Did the companies prosper? These are among the questions that the limited historiography of the survey contractors has consistently pointed to over the years but has never attempted to answer.

THE LEGAL AND REGULATORY FRAMEWORK

Government-issued authorizations[7] to conduct surveys of public land were issued under two separate statutes, that of May 31, 1875, and later under the law of December 15, 1883, which replaced the 1875 statute. Both laws stated as their principal objective the "colonization" of Mexico's lightly populated hinterlands. Very few surveys were authorized under the brief, two-sentence provision of the 1875 law, which empowered the president to appoint "exploratory commissions . . . in order to obtain colonizable land with the requisites necessary for measurement, appraisal and description." The "commissions" were to be compensated with "one third of the said land or of its value."[8] The 1883 law, passed during the administration of President Manuel González, controlled virtually all survey activity until 1902, when the use of private companies to survey public land was prohibited by presidential decree. Like the 1875 law, Article I of the 1883 legislation expressly linked the surveys to colonization: "In order to obtain the land necessary for the establishment of colonists, the Executive will order the survey, measurement, division and appraisal of the vacant land or national property that may exist in the Republic, appointing to that end commissions of engineers that it may consider necessary, and determining the system of operations that will be followed."[9]

The president was empowered to authorize companies to survey the

baldíos and to transport and establish colonists on the land. Once authorized to survey in a particular state or substate jurisdiction (commonly a
district), the companies had to obtain a second authorization to begin the
survey from the federal district judge in whose jurisdiction the baldío
could be found. If no opposition to the survey was expressed, the judicial
record of the proceedings would be forwarded to the Secretaría de Fomento. If someone opposed the survey, the appropriate *juicio* (judicial
proceeding) would take place. In compensation for the surveys, the companies would be paid "one-third of the land that they survey, or of its
value." No limits on the size or the use of a company's share were imposed by the law.[10]

The heyday of the survey business occurred in the late 1880s. José L.
Cossío's nationwide tabulation of survey contracts, as reported in Tables
4 and 5, indicates that the earliest contracts were let in 1881, when four
were signed. (In the six states that make up this study's sample, however,
I found two issued in 1879.) Cossío claims there were twenty contracts
before 1884 (and therefore probably issued under the authority of the
1875 law), or about 12 percent of his total of 169. He reports that all but
one of them were issued between 1881 and 1891, with the majority of
those issued in just one four-year span, 1886–90.

In the six-state study area, authorizations to survey land were issued to
fifty-nine individuals or companies, invariably in response to a letter that
requested an authorization and also specified the states, or the state or
substate jurisdictions, that the prospective contractor sought to survey.
These requests were rarely denied. Evidence of only nine denials in the
study area could be found, indicating an acceptance rate of nearly nine
out of every ten requests.[11] Apparently, almost anyone who asked for a
survey contract got one. So many authorizations were being issued by the
early 1880s that Fomento used a preprinted form to accept the applica-
tions for authorization, merely filling in a few blanks to indicate the
name of the contractor, the area to be surveyed, and the dates. The language varied little from one contract to another; variations related mainly
to authorizations that went beyond the survey of baldíos alone to include
the survey of demasías as well, the power to arrange composiciones with
landholders, the right to buy part of the government's two-thirds share at
a specified price, and the obligation to establish colonies on surveyed
land. Contracts were occasionally canceled by the government, most frequently because of the contractor's failure to begin the survey within the
three-month limit usually specified, but sometimes because of negligence
or incompetence.

Very few contracts actually required the surveyor to colonize the land,
and those contracts that did tended to provoke protests by the surveyors

Table 4   Authorizations to Survey in Mexico, 1876–1911

Recipient	Jurisdiction	Year
José Matilde Alcocer	Yucatán	1883
I. Altamirano y J. F. Bulman	Tehuantepec, Oaxaca	1883
Guillermo Andrade	Puerto Isabel, B.C.	1889
Felipe Arellano y Socios	Tepic y Zacatecas	1888
F. Armendaiz y R. García Martínez	S.L.P., Querétaro	1887
Rafael Arrillaga y Eduardo Paez	Veracruz	1891
F. Arteaga	Guerrero	1887
Antonio Asúnsolo y Cía.	Chihuahua, Durango	1884
Antonio Asúnsolo	Chihuahua	1887
José María Becerra	Sinaloa	1887
Gen. Juan C. Bonilla, Nicolás Islas y Bustamante, Miguel R. Méndez	Puebla	1883
Juan Bottero	Sultepec, Mina in México y Guerrero	1890
Samuel Brannan	Sonora (entre Comoripa y Sahuaripa)	1883
Adolfo Bulle	Santa Gertrudis y San Borja, B.C.	1884
Bulnes Hermanos	Tabasco, Chiapas	1883
Bulnes Hermanos	Chiapas, Durango	1887
Manuel Bulnes y Socios	Guerrero	1886
Manuel Bulnes y Socios	Tabasco	1888
Manuel Bulnes y Socios	Veracruz	1886
Manuel Bulnes y Socios	Autlán, Mascota, Lagos, La Barca, Jalisco	1888
Manuel Bulnes y Socios	Zapotlan y Sayula, Jalisco	1888
M. Bulnes	Oaxaca	1886
Juan Bustamante	Haciendas El Sotol, La Carbonera, S.L.P; Nuevo León, Coahuila, Zacatecas	1889
Pedro Bustamante	Puebla	1890
Pedro Bustamante y José Iglesias	Querétaro y Tamaulipas	1888
Emiliano Busto	Guanajuato	1887
Gen. Juan B. Caamano	Rail right of way, México, Michoacán, Guerrero	1890
B. O. Cagigas	Michoacán	1886
José M. Calderón	Durango	1889
Fernando Calderón	Zacatecas	1885
José Cárdenas	Guerrero	1890
José Cárdenas	Tabasco	1890
Lauro Carrillo	Arizpe y Sahuaripa, Sonora	1886
Joaquín D. Casasús	Durango	1887
Manuel Castro	Dist. Sur, Sonora	1881
José Ceballos Muñoz y Cía	Muzquiz, Coahuila	1887

Table 4—Continued

Recipient	Jurisdiction	Year
José Ceballos	Monclova, Coahuila	1891
Luis Ceballos y Cía.	Chihuahua, Durango, Sonora, Sinaloa,	1886
Celada Hmnos. y R. Garma	Michoacán	1890
Jesús María Cerda	Lampazos de Naranjo, Nuevo León; Río Grande, Coahuila	1888
Hipólito Charles	Sonora	1889
Ignacio T. Chavez	Aguasc.	1890
Cía. Colon. Mexicana de New York	Chiapas	1881
Cía. del FFCC y Teleg. de Texas, Topolobambo, y Pacífico	Sinaloa, Sonora, Chihuahua, Coahuila	1886
Cía. El Progreso	B.C. (2 munic.)	1889
Cía. Ltd. del FFCC Central Mexicano	Tamaulipas	1890
Cía. Zacatecana	Zacatecas, Aguasc., Durango, S.L.P.	1883
Mariano Coronado	6th, 8th, 9th, 10th cantones, Jalisco	1887
Vicente Dardon y Socios	Tabasco	1889
Juan M. Davalos	Guerrero	1891
Emeterio De la Garza	Tamaulipas	1888
Guadencio G. De la Llave, José M. De la Vega y Socios	Puebla (3 munic.)	1887
F. L. De Saldana y Cía.	Tamaulipas	1886
J. Dublan y Cía.	Parras, Coahuila	1886
C. Eisenman	B.C.	1886
Manuel Escobar Escoffie y Socios	Durango	1889
Gens. Ignacio Escudero y Epifano Reyes	Lago Cuitzeo, Michoacán	1891
Ramon Fernández	S.L.P.	1883
Ramon Fernández	S.L.P. (6 munic.)	1884
Jaime Ferrer	Durango	1889
Gen. Alonso Flores y Castulo Zenteno	Tamaulipas	1882
Gen. Alonso Flores y Fsco. Poceros	Tamaulipas	1883
Conrado Flores y Santiago H. Hale	B.C.	1883
F. Gallastegui y J. D. Nava	Cuencamé, Durango	1887
Rafael García Martínez y Socios	Durango, Sinaloa, Jalisco, Zacatecas	1887
Rafael García Martínez y Socios	Michoacán	1884
Rafael García Martínez	S.L.P., Querétaro	1888
Rafael García Martínez	Sinaloa	1889
Estanislao García Mora	10th cantón, Jalisco	1888
Luis García Teruel	Puebla, Tlaxcola, Michoacán, S.L.P.	1885
Maríano García	Durango	1889
Antonio Gayon y José Patricio Nicoli	Jalpan, Querétaro	1886
R. Gibert	B.C.	1887

Table 4—Continued

Recipient	Jurisdiction	Year
P. Gómez del Campo	Chihuahua	1883
Ignacio Gómez del Campo	Sonora (entre Ríos Yaqui y Mayo)	1883
Ramon Gómez y Pena	2d cantón, Jalisco	1886
Jesús González Treviño e Hijos	Río Grande, Coahuila	1891
Joaquín María González	S.L.P.	1889
Agustín Guerrero	Parras, Viezca, San Pedro, Coahuila	1889
Andrés Gutt	Chiapas	1886
Jesús Guzman y Braulio Sánchez	Michoacán	1883
Nieves Hernández	Tamaulipas	1885
José M. Herrera	Querétaro	1886
Gen. Pedro Hinojosa	4th Dist., Nuevo León	1891
T. Hinojosa y S. Seguin	Tamaulipas	1887
J. Iñigo	La Bajada, Chihuahua	1886
A. M. Ituarte	Allende, Guerrero	1889
José Manuel Jofre y Vicente Ordozgoiti	Veracruz (2 munic.)	1891
E. Kosterlitzky	Los Horiones, Sonora	1888
Pedro Landazuri	Jalisco	1883
Donaciano Lara y Manuel Santibanez	Veracruz	1887
J. F. Lopez	Aguasc.	1886
Pablo Macedo	B.C.	1886
Hesiquio Maranon	Veracruz	1885
Luis Martínez de Castro	Sinaloa, Chiapas	1899
Luis Martínez de Castro	Sinaloa, Chiapas	1901
L. Martínez Urista y Carlos Eisenman	S.L.P.	1888
Faustino Martínez y Cía.	Yucatán	1890
Gen. Ángel Martínez y Pedro Landazuri	Colima	1883
Gen. Ángel Martínez y Francisco C. Palencia	Colima	1888
Federico Méndez Rivas	Tabasco, Chiapas	1883
Federico Méndez Rivas	Tabasco	1886
Federico Méndez Rivas	Veracruz	1886
Miguel R. Méndez	Puebla	1884
Rafael Mendoza	Monclova, Río Grande, Coahuila	1890
Rafael Mendoza	Tabasco	1889
Justiano Mondragón	Veracruz	1889
Francisco Monsalve	Nuevo León	1888
José Mora	Pichucalco, Chiapas	1890
José Rafael Mora	Michoacán	1890
Sabino Mugia y Lagarde	Tepeji, Puebla	1891
Victoriano Muñoz	Union, Galeana, Guerrero	1889
Eduardo Noriega	Guanajuato	1890

Table 4—Continued

Recipient	Jurisdiction	Year
Marcial Nuñez	Tamazula, Durango	1890
Manuel Olaguibel	Matamoros, Puebla	1889
Franciso Olivares	Ures y Altar, Sonora	1887
Francisco Olivares	Alamos y Hermosillo, Sonora	1889
Manuel Orellana Nogueras y Cía.	Guanajuato	1884
Plutarco Ornelas	Tamaulipas, Coahuila, Chihuahua	1881
Agustín Ortega	Campeche	1889
Albert K. Owen	Rail right of way in Sinaloa, Sonora, Chihuahua, Coahuila	1890
Enrique Pacheco y Ignacio Sandoval	Chihuahua y Sinaloa	1890
C. Para	Mascota, Jalisco	1886
Ireneo Paz	B.C.	1884
Rosendo Pineda	Chiapas	1885
Carlos Quaglia y Cía.	Cadereyta y Tolimán	NA
Carlos Quaglia	Parras, Viezca, San Pedro, Coahuila	1888
Carlos Quaglia	Río Grande, Coahuila	1891
Carlos Quaglia y L. García Teruel	Sonora, Oaxaca (Jecker)	1885
Antonio V. Quiroz	Veracruz (3 munic.)	1887
Manuel Ramírez Varela	Jamiltepec, Oaxaca	1886
Leóncio Rigo	Guadalajara, Teocaltiche, Tequila Ocotlan, Jalisco	1890
Benjamín Ríos	Veracruz	1884
Carlos P. Rivas	Puebla	1891
Carlos P. Rivas	Veracruz	1891
M. Roca	Campeche	1887
E. A. Roller y Cía.	Tamaulipas	1886
Demetrio Salazar y Encarnación Davila	Coahuila	1885
Demetrio Salazar	Rancho San Carlos, Durango	1886
Demetrio Salazar y Braulio Sanchez	Michoacán	1886
Felipe Salazar	Nuevo León, Tamaulipas	1886
Refugio Salazar	Alamos, Sonora	1889
Manuel Sánchez Mármol	Tabasco	1887
Ignacio Sandoval	Mina, Chihuahua	1887
M. Santabanez, M. C. de la Lastra y Jesús Urias	Guerrero	1887
M. Santibanez y M. C. de la Lastra y Cía.	Guerrero	1886
Manuel Santibanez y Aristeo Mercado	México (8 munic.)	1889
M. Santibanez, A. Pradillo, G. Petriz, M. Moreno, M. D. Santibanez	Oaxaca	1886
L. Santos y B. Rodríguez	Río Grande, Monclova, Viezca, Coahuila	1888

Table 4—Continued

Recipient	Jurisdiction	Year
Fernando Sayago	Jalisco	1888
Jesús Serrano	Tlaxcala	1890
Antonio Tovar	Parras, Viezca, San Pedro, Coahuila	1889
Antonio Tovar	Saltillo, Coahuila y Mapimi, Durango	1890
Antonio Tovar y Pedro Bustamante	Puebla	1890
Jesús Trevino	Nuevo León	1888
Rodolfo Valdéz Quevedo	Galeana, Chihuahua	1888
Eduardo Valdéz y Cía.	Sonora y Sinaloa	1885
José Valenzuela y Fernando Zetina	Hidalgo	1890
P. Valenzuela	Sierra de la Parida in Zacateca y Jalisco	1888
José Valenzuela	Choix, Sinaloa	1891
Manuel Vallejo	Michoacán	1886
José María Velazquez	Veracruz	1889
Manuel S. Vila	Campeche	1886
Manuel S. Vila	Jalisco	1886
Manuel S. Vila	Jalisco	1890
Rafael y Javier Villaurrutia	Hidalgo	1887
Eduardo Clay Wise y Socios	Chiapas	1881
Juan Manuel Zambrano	Parras, Viezca y San Pedro, Coahuila	1888
Total Contracts	169	

Source: José L. Cossío, *¿Cómo y por quiénes se ha monopolizado la propiedad rústica en México?* (Mexico: Editorial Jus, 1966), pp. 78–94.

Note: "Aguasc." is short for Aguascalientes, "B.C." is short for Baja California, and "S.L.P." is short for San Luis Potosí.

that colonization, at least on the terms envisioned by the government, was simply unworkable in Mexico. Luis Martínez de Castro's survey contract of 1899 for Sinaloa and Chiapas required him to settle two families for every one thousand hectares he acquired, in the proportion of three European families to one Mexican. When Fomento increased the requirement to four families for every one thousand hectares, Martínez de Castro argued that the colonization requirements were unrealistic, "the colonization of Mexico being in itself so difficult." In 1901 the contract was replaced with one that made no references to colonization.[12] The failure of the Land Company of Chiapas to fulfill its colonization requirements is treated in chapter 3. A company official concluded that the firm's experience with colonization proved that "the installation of colonists

Table 5 Survey Authorizations
by Year

Year	Number
1881	4
1882	1
1883	15
1884	8
1885	8
1886	31
1887	22
1888	20
1889	23
1890	22
1891	12
1892	0
1893	0
1894	0
1895	0
1896	0
1897	0
1898	0
1899	1
1900	0
1901	1
1902	0
1903	0
1904	0
1905	0
1906	0
1907	0
1908	0
1909	0
1910	0
1911	0
Total	169

Source: Table 4.

is totally impossible" in the proportion and nationality fixed by the government in its contract. He estimated that the company had spent 325,000 pounds sterling to settle colonists, money that had not been recovered by the sale of land. The most feasible approach to colonization was not to set a strict timetable with quotas of colonists, but to undertake it "little by little," establishing "profitable agricultural enterprises and plantations of rubber, coffee, cacao, etc. that would provide permanent work to the inhabitants of the region and to new settlers."[13]

By the mid-1890s it was becoming clearer that the country's decades-old faith in colonization—particularly by European immigrants—as one of the principal sources of the country's development had been badly misplaced. "This undertaking to colonize our land," wrote the nineteenth-century legal scholar and agrarian theorist Wistano Luis Orozco in 1895, "probably obeys, as the principal and most powerful motive, a puerile spirit of imitation of the United States."[14] Like all the previous legislation designed to encourage colonization by foreigners, the colonization provisions of the laws of 1875 and 1883 were ineffective.[15] In 1900 Fomento admitted that the public lands in remote regions were not "propitious" for the settlement of foreign colonists, and two years later it published a frank, 104-page analysis of the failure of the regime's colonization schemes; by then, a mere 7,962 colonists were residing in thirty-two settlements.[16]

In the following pages I will lead the reader through a typical survey, emphasizing the legal and procedural framework in which the companies, the federal and local authorities, and landholders were expected to interact. I will offer generalizations about methods of operation that are based on my reading of the archival record. Since every record of every survey conducted in the six-state study area was examined, the data presented here and in the chapters that follow are exhaustive for those six states, unless otherwise noted.

Most of the survey companies were charged only with surveying any baldíos (and oftentimes demasías as well) they could find in the area specified in the contract. The contracts required them to begin work within three months, to finish within three or five years, and to bear all the costs of the "survey, demarcation, subdivision of land and the drawing up of the corresponding maps," which were to be sent to Fomento along with the judicial proceedings. Prior to beginning a survey, companies first had to designate as precisely as possible, before the appropriate federal district judge, the location of the baldíos they wanted to survey. To discourage the companies from upsetting "unnecessarily the owners of the land," Fomento insisted in 1883 that when the companies sought a judge's authorization, they make their designations of land to be surveyed "with positive evidence of the existence" of baldíos.[17]

During this process and until all the documents were returned to Fomento for its approval, the survey was generally out of Fomento's control and in the hands of the judiciary, although judges, survey companies, and aggrieved property holders wrote constantly to the ministry for advice and assistance. Still, the companies were not required to file regular reports on their activities; their freedom to survey whatever land suited their interests, without regard to the total amount of baldío available for survey (as long as it limited itself to the area authorized in the contract), was recognized by Fomento.[18]

Surveys were usually supervised by a district or municipal judge, under the direction of a federal judge.[19] Property owners in the affected zones were instructed to present—at a certain location on a specific date—their titles or other documents proving ownership. Notification was usually supervised by the local judge, who frequently accompanied the surveying parties. The charge to the local judge by the federal judge often contained language such as the following, requiring him to proceed

> at all times with prior summonses to the adjacent property holders, in person to those who are present and by requisition to those who are absent, informing them in advance of the day, time and place that the survey should commence. The commissioned Judge will take care that the above-mentioned persons do not occupy in any way the land being measured until such time as they may acquire the property legally; in cases to the contrary they will be punished as squatters on national property.[20]

The following citation to attend a survey was issued to local inhabitants by a federal judge in Sinaloa in 1891:

> To the owners or possessors of the property known as El Salado: Being required to carry out the measurement of "Caitime," which abuts this rancho, I beg you to be kind enough to attend with the titles or documents of the property on the fifth day of this October, at ten in the morning, or attend the measurement of the lines that correspond to you and defend your property lines, being kind enough to return this communication signed to indicate your acknowledgment of it; advising you that in the event of your failure to do so, you will be subject to the damages prescribed by law. S/Alejo Castro.[21]

When the individuals cited appeared at the designated place, documents were presented and the survey proceeded, with the local judge entering a record of events. The following partial record is from the survey of the *municipios* (municipalities) of Nonoava and Sisoguiche in Cantón Abasolo, Chihuahua, by the firm of Ramón Guerrero and Ignacio Gómez del Campo, in 1883:

> Today being the 21st of January of 1883, in the area and property of the Municipality of Nonoava, designated as colonizable by Lic. Guillermo Urrutia, general agent of the survey Company, Gómez del Campo, Guerrero, Macmanus and associates, and in the point called Lagunita de Chupachic on the mesa of the same name, being present the above-mentioned gentleman, the headmen of the Pueblo of San José Baguiachic, Ramón and José, who are without surnames, the witnesses in attendance and the Survey Engineer C. Primitivo Saenz, the judge of record prepared and proceeding in accord with the law to the operations related to his assignment, by virtue of which and situated on the S.E. bank of the above-mentioned Lagunita, here was placed the marker and the flag; and the Engineer having indicated that for the operations the method of triangulation and geometric positions would be appropriate, it being required by the topographical situation of the land, he proceeded immediately to lay out his base in the direction of 72 [degrees] North and 45 [minutes] East, and with a well-accepted steel tape of twenty meters, he measured the distance of 2,240 meters. . . . He concluded the present proceeding on the 14th of February of the year mentioned, being in conformity all the adjoining property owners, as can be seen in his writ of the 9th of the same month that is attached, those in attendance having signed with the Judge. To this we bear witness.[22]

In his survey of the *directoria* (directory) of Choix in Sinaloa's Fuerte district, José Valenzuela designated seventeen lots for survey before the federal judge. The *alcalde* (mayor) of Choix supervised the survey, in which the holders of a parcel known as Toipaque were cited to appear. They presented a title issued in Guadalajara in 1766:

> In the rancho of Toipaque on the 2nd day of February of 1892, the court being installed in the house of Doña Jesús Usárraga, widow of Martínez, and being present the Messrs . . . , all possessors of the properties that are going to be surveyed, Román Aragón presented the title of Toipaque. . . . Finishing the reading and recording of the mentioned documents, the expert appointed for the survey, who

was present, declared that he had surveyed some months ago the properties of . . . which are adjacent to those which they are now going to measure and that he knows therefore the boundaries of this part, which are as follows: . . . Those present being in agreement with the indicated boundaries, the hearing was considered done and the next day was set aside to verify the rest of the boundaries, beginning from the marker of the San José mesa, where those who have been summoned will meet at eight in the morning. With this arrangement the present proceeding, which I sign with those who know and with witnesses in attendance, is closed.[23]

Typically, the judicial record of a survey consisted of a compilation of dozens of such entries, documenting the notification of property holders, their appearance before the judge as the surveyors progressed through the territory to be measured, evaluation of the documents by the judge or survey company, the placement of markers indicating property lines, and the property holders' statement of either opposition or "conformidad" with the placement of the markers. Inhabitants of the area usually assented to the location of the markers; if they did not, a hearing was held before the federal judge, who ruled on the validity of the property holder's claim. After the local judge sent the papers to the federal judge, the latter ordered the publication of notices in the state government's official newspaper, three times in thirty days, as well as their posting in public places. Anyone who considered their rights infringed by the survey had thirty days after the last publication date to appear before the judge. When opposition surfaced, it was almost always because a property holder believed that land declared baldío by a company was in fact property titled to him.

The surveyor was also required to file a report on the physical and economic geography of the area and a technical or geometric description of the land surveyed, as well as to estimate the value of the land and propose a division into three zones, one of which would be indicated as the company's preferred compensation.

After the survey was completed and the district court had fulfilled its obligations, the entire file was sent to Fomento, where officials reviewed it for compliance with the procedural formalities. Judging by the memorandums produced by the reviewers, their main interest was in determining whether opposition to the survey had appeared and whether it had been resolved. If opposition was still pending, the ministry would sometimes rule on the validity of a claim itself, bypassing the judiciary, or mediate a settlement between the company and the property holder.

Before the title to the company's one-third share could be issued, however, Fomento also had to decide whether to accept the company's proposed division of the land, which it usually did. The surveyor's report was its main guide to the value of the one-third as well as the rest of the surveyed land.[24] The surveyor generally stated that the government's share equaled the value of the portion requested by the company. For example, in Cantón Balleza, Chihuahua, Ignacio Gómez del Campo's firm divided its 685,717 hectares of surveyed baldío into twelve zones, ranging in area from 29,757 hectares to 144,810 hectares. The company requested title to four zones totaling 225,246 hectares, or 3,326 hectares less than one-third of the total. The surveyor assured Fomento that "the division of the zones is equitable; they are equal from any point of view that may be considered," estimating the value of all the land at twenty centavos a hectare. After receiving the file, a Fomento functionary reported to the secretary, General Pacheco, that the four zones requested by the company were in the most populated area of the cantón. He added that a railroad passed through one of the zones linking it with the mines of the Sierra Madre, and that part of the San Juan River passed through another zone. The functionary further indicated that the eight zones reserved for the government appeared to contain equivalent amounts of water and good land, judging by the surveyor's report. His only proposed revision was to raise the surveyor's estimate of the land's value from twenty to fifty-five cents a hectare. Fomento approved the company's request, but on the condition that the land be returned in the event any of it was reclaimed by individuals with "legitimate rights." The title was issued to the company in July 1885, four months after the file was sent to Fomento.[25]

The division of surveyed land became a contentious issue if the government decided that the property being reserved for it was clearly inferior or inconveniently located. When Jesús Valenzuela's company identified thirty-six separate lots of baldío totaling 241,600 hectares in Chihuahua's Cantón Iturbide, the firm's representative, José Valenzuela, complained that the company had lost money on the survey, and that the baldío it found was completely lacking in "elementos." So he offered to let the government keep all 241,600 hectares, provided the company could take its one-third in compensation from one lot in Cantón Ojinaga. This offer was quickly rejected by President Díaz because, Fomento told Valenzuela, "it is not convenient for the government to have many pieces of separate lots." Fomento turned the tables on Valenzuela, deciding that he would have to take all thirty-six lots in Iturbide, while the government would receive the equivalent of its two-thirds by taking back land it had previously titled to the company in zones contiguous to property the government already held elsewhere in Chihuahua.[26]

When Fomento questioned the fairness of the same company's proposed division in Cantón Camargo, José Valenzuela argued that the one-third requested by the company was actually the least valuable part of the land it surveyed because it was the farthest from any settlements and contained less water and timber. Still, he asserted, the division he proposed was the most equitable possible. Fomento accepted his reasoning, probably not without some skepticism that a contractor would choose the least valuable land for himself.[27]

It took Fomento and the Mexican Land Colonization Company six years to agree on the division of land in Sonora's Magdalena district, a total of 474,449 hectares of baldío divided into numerous zones. The company's proposal was countered by one that would deprive the company of lots along the U.S. border and along a railroad right-of-way. Emilio Velasco, the company's lawyer, rejected the offer because it gave "all the advantages to the Government." They compromised by splitting one of the best lots.[28] The division of land in Tabasco proposed by Policarpo Valenzuela was rejected because the lots reserved for the government were low and swampy, while the surveyor wanted to keep land that, if distributed in discrete lots, was of higher quality.[29]

Judging by the archival record, many surveys were conducted with care and attention to detail, whereas the reports of others were clearly perfunctory and suggested a corresponding lack of diligence. Occasionally, judicial authorities seemed to be entirely absent from the survey party. When no local judge was available an alcalde, acting as a town magistrate, supervised the survey within the borders of that municipality. Sometimes neighboring property holders were not cited at all. In Rafael García Martínez's survey in Durango's Mezquital *partido* (district), for example, the surveyors told the partido's local judge that the inhabitants of the pueblos of Jaconoxtle, Taxicarniga, Santa María de Ocotán, San Francisco, Teneraca, San Buenaventura, San Lucas, and San Antonio were not cited to attend the survey, in part because none of the pueblos had a *fundo legal* that might be threatened by the survey and because "they are in their totality pure Indians and have no governing authorities" save for "gobernadores" who keep public order under the general supervision of the jefe político of Mezquital and the municipality of Huazamota. Nevertheless, the company set aside as *ejidos* four *sitios de ganado mayor* (7,024 hectares) for each village.[30]

Startling omissions in the judicial record were sometimes noted by Fomento. For example, in the Becerra Hermanos' survey in Sinaloa's Fuerte district in 1891, a ministry official noted in a memorandum the absence of opposition, even though neighboring property holders "did not sign the description of property lines." Notwithstanding this oversight, four days after the memo was written President Díaz signed a title giving the company its one-third share.[31] In Tabasco, the surveys of the partidos of Jonuta and Balancán were conducted "without the intervention of any judicial authority," Fomento noted. The federal judge merely appointed two individuals to accompany the survey to witness the signings by property holders.[32] Three years later, a judge in the same state admitted he could not tell Fomento whether the survey by Policarpo Valenzuela was being carried out "with all the legal requisites."[33] Such procedural errors and omissions were notable for their infrequency, however, judging by the archival records of the proceedings.

THE DISTRIBUTION OF CONTRACTS AND COMPENSATORY LAND

In the six states selected for this study, Fomento's records indicate that fifty-nine concessions to survey public lands in Chihuahua, Durango, Sonora, Sinaloa, Chiapas, and Tabasco were awarded before 1911. All of them were issued between 1879 and 1901, and all but three before 1891. Thus the selection of the contractors, like the surveys themselves, was largely confined to the 1880s. One-third of the fifty-nine concessions was awarded during the presidency of Manuel González.

The concessions were actually issued to just forty-nine different business organizations, as some held more than one, as Table 6 indicates. Of the fifty-nine concessions, nineteen were transferred to other firms with Fomento's permission, as shown in Table 7 and Table 8. Twelve of the seventeen firms that acquired these transferred concessions never received an original concession from the government; they entered the business only as buyers of another holder's right to survey. One contractor, Luis Huller, ended up with nine concessions, all of which eventually were acquired by the British-owned Mexican Land Colonization Company.

The two most striking patterns are the high proportion of concessions that were awarded but never implemented, and the low proportion of active survey companies that acquired any land in compensation for their labors (Table 9). In twenty of the original fifty-nine concessions, no land was ever surveyed and the concession itself was never transferred. At least one out of three concessions, therefore, was never activated. Of the sixty-one firms that held at least one concession, only twenty-six ever acquired land in compensation for a survey.[34] Others did survey work, but it was either rejected by the government, never finished, or finished by one of the twenty-six when the concession was transferred.

Both of these patterns point to a surprisingly low success rate among survey contractors. Individuals who did little or no work and then

Table 6 Survey Contracts Issued by Fomento
in Six-State Study Area

Contractor	Year	States	T = transfer F = foreign[a]
Guillermo Andrade	1889	Sonora	—
Antonio Asúnsolo	1884	Chihuahua, Durango	—
José María Becerra	1886	Sinaloa	T
Adolfo Bulle[b]	1884	Sonora	T
Bulnes Hermanos	1884	Chiapas, Tabasco	—
Bulnes Hermanos	1884	Chiapas, Tabasco	—
Bulnes Hermanos	1888	Tabasco	—
José María Calderón	1889	Durango	T
José Cárdenas[b]	1889	Tabasco	—
Lauro Carrillo	1888	Sonora	—
Joaquín Casasús	1887	Durango	—
Luis Ceballos[b]	1886	Chihuahua, Sonora, Sinaloa, Durango	—
Hipólito Charles	1889	Sonora	—
Carlos Conant[b]	1890	Sonora	T
Rafael Cruz[b]	1881	Sonora	—
Vicente Dardon[b]	1889	Tabasco	—
Jesús Díaz González[b]	1883	Chihuahua	T
F. Gallestegui & J. Nava[b]	1887	Durango	—
Maríano Gallego	1883	Sinaloa	T
Telésforo García[b]	1883	Durango	—
Telésforo García[b]	1883	Sonora	T
Telésforo García[b]	1883	Sinaloa	T
Telésforo García[b]	1883	Sinaloa	T
T. García & Jesús Meraz[b]	1883	Durango	—
Rafael García Martínez	1883	Durango, Sinaloa	F
José María Garza Galan[b]	1879	Chihuahua	—
Ignacio Gómez del Campo[b]	1883	Sonora	T
I. Gómez del Campo & Ramón Guerrero	1882	Chihuahua	—
Patricio Gómez del Campo, A. Asúnsolo & Macmanus[b]	1883	Chihuahua	—
Andrés Gutt[b]	1886	Chiapas	T
Luis Huller	1886	Sonora	T,F
José Inigo	1886	Chihuahua	—
Manuel Jamet & Jaime Sastre[b]	1883	Tabasco	—
Luis Martínez de Castro	1899	Sinaloa, Chiapas	—
Luis Martínez de Castro	1901	Sinaloa, Chiapas	F
Federico Méndez Rivas[b]	1883	Chiapas, Tabasco	T

Table 6—Continued

Contractor	Year	States	T = transfer F = foreign[a]
Federico Méndez Rivas[b]	1886	Tabasco	—
Meza Hermanos	1882	Durango	—
Juan B. Ochoa[b]	1879	Chihuahua, Durango	—
Francisco Olivares	1887	Sonora	F
Francisco Olivares[b]	1889	Sonora	—
Plutarco Ornelas	1883	Sonora, Durango	—
Manuel Peniche	1885	Sonora	T, F
Manuel Peniche	1885	Sonora	T, F
Rosendo Pineda[b]	1885	Chiapas	—
Carlos Quaglia & Luis García Teruel[b]	1885	Sonora	T
Andrés Quijano[b]	1890	Sonora	—
Demetrio Salazar[b]	1886	Durango	—
Demetrio Salazar[b]	1885	Durango	—
Refugio Salazar[b]	1889	Sonora	—
Manuel Sánchez Mármol[b]	1887	Tabasco	T
Ignacio Sandoval	1886	Chihuahua	—
I. Sandoval & Enrique Pacheco	1890	Chihuahua, Sinaloa	—
Manuel Thomas y Teran[b]	1885	Tabasco	T
Antonio Tovar[b]	1890	Durango	—
Eduardo Valdéz[b]	1885	Sonora, Sinaloa	T
Rodolfo Valdéz Quevedo[b]	1888	Chihuahua	—
Jesús Valenzuela et al.	1882	Chihuahua	—
José Valenzuela	1891	Sinaloa	—

Total Contracts: 59

Source: Archivo de Terrenos Nacionales, Sec. de Reforma Agraria.

Notes: In those concessions in which the owners clearly maintained shares in the new enterprise, the concessions are not counted as transfers.

[a] "Transfer" means the concession holder transferred it to another party. "Foreign" means there was some foreign investment in concession; the proportion varied from a minority share to all shares.

[b] There is no evidence that the contractor carried out any surveys under this contract.

transferred their concessions probably did not command a very high price for them, since it was apparently not difficult to acquire one directly from Fomento. Requests for survey concessions were turned down in only six cases, with the government encouraging competition by issuing concessions that overlapped in both time and space, an issue treated in

Table 7 Contractors Who Acquired Concessions by Transfer

Name	Number	Foreign
Bulnes Hermanos	1	—
Compañía de Irrigación de Sonora	1	—
Telésforo García et al.	2	—
Mariano García	1	—
Luis Gayou	1	—
Luis Huller	8	F
International Co.	9	F
Land Co. of Chiapas	1	F
Tomás Macmanus	1	—
Manuel Martínez del Río	2	—
Mexican Land Colonization Co.	9	F
William T. Robertson	1	F
Domingo Rosellon	1	—
Simon Sarlat	1	—
Natalia Valenzuela de Saracho	1	—
Jesús Valenzuela et al.	1	—
Policarpo Valenzuela	1	—

Source: Archivo de Terrenos Nacionales, Sec. de Reforma Agraria.

chapter 3. Those who did some work but sold their concessions before submitting the survey for Fomento's approval cannot be completely distinguished from those who acquired concessions merely to sell them. In any event, the archival record indicates that a survey concession did not guarantee success as a land developer. Only two out of five concession holders ever received land in compensation, and it seems unlikely that those who sold their concessions were able to charge much for them. Most survey contractors therefore never received any compensation—either they did nothing, their work was deficient and was therefore not compensated by Fomento, or they were unable to generate the cash flow necessary to cover their expenses until they received a title.

Foreign capital's participation in the survey business cannot be absolutely quantified. No archival source routinely identified even the names of company investors, let alone their nationality. Thus the presence of foreign investors could only be detected when their existence was specifically referred to. Firms identified in Table 6 and Table 7 as carrying foreign investors are only those for which corroborating evidence was

Table 8 Survey Contract Transfers

Seller	Buyer 1	Buyer 2	Buyer 3	Buyer 4
Becerra	Luis Gayou	Robertson	Valenzuela de Saracho	Martínez del Río
Bulle	Int. Co.	Land Colon. Co.	—	—
Calderón	Rosellon	—	—	—
Conant	Cía. de Irrigación	Macmanus	—	—
Díaz González	Telésforo García	J. Valenzuela	—	—
Gallego	Telésforo García	L. Huller	—	—
Telésforo García	L. Huller	Int. Co.	Land Colon. Co.	—
Telésforo García	L. Huller	Int. Co.	Land Colon. Co.	—
Telésforo García	L. Huller	Int. Co.	Land Colon. Co.	—
Gómez del Campo	L. Huller	Int. Co.	Land Colon. Co.	—
A. Gutt	L. Huller	Int. Co.	Land Colon. Co.	Chiapas Land
L. Huller	Int. Co.	Land Colon. Co.	—	—
Méndez Rivas	Bulnes	—	—	—
Quaglia and Teruel	L. Huller	Int. Co.	Land Colon. Co.	Martínez del Río
Sánchez Mármol	P. Valenzuela	—	—	—
Thomas y Teran	Sarlat	—	—	—
Valdéz	M. García	L. Huller	Int. Co.	Land Colon.

Source: Archivo de Terrenos Nacionales, Sec. de Reforma Agraria.

found. Not even the proportion of foreign participation could be determined in most cases. The designations of "foreign participation" among the firms in these tables are therefore not intended to be categorical, and the interpretations that are based on them must be considered tentative. With that caveat in mind, two patterns of foreign participation emerge from the archival record. The first relates to the acquisition of contracts, and the second to compensation.

In only six of the fifty-nine original concessions was there evidence in the archival records of investments by foreigners; only one concession was awarded outright to a foreigner, Luis Huller.[35] Yet foreigners would later acquire an additional ten concessions; Huller alone acquired eight of the ten, passing them on to his successor companies. Thus foreigners tended to enter the business not by requesting a contract from Fomento but by buying concessions from other companies.

Of the twenty-six contractors that received land in compensation, six were either totally foreign or were dominated by foreigners.[36] This suggests a rather low level of activity by foreigners, but the number is mis-

Table 9 Ranking of Survey Companies in Six States by Land Titled in Compensation

Contractor	Area Titled (ha)
Jesús Valenzuela y Socios	3,001,551
Francisco Olivares[a]	1,648,395
Ignacio Gómez del Campo y Ramón Guerrero	1,268,983
Mexican Land Colonization Co.[a]	958,139
Manuel Peniche[a]	651,034
Ignacio Sandoval	633,528
Antonio Asúnsolo	596,364
Rafael García Martínez[a]	563,954
Joaquín Casasús	207,608
Policarpo Valenzuela	204,944
Luis Martínez de Castro[a]	185,400
Sinaloa Land Co.[a]	104,923
José María Calderón	95,680
Luis Huller[a]	94,665
Lauro Carrillo	78,332
International Co.	55,982
Jesús Díaz González	22,920
Plutarco Ornelas	20,284
Mariano Gallego	18,687
Ignacio Sandoval & E. Pacheco	17,984
José Inigo	14,835
José María Becerra	12,049
Hipólito Charles	10,283
Natalia Valenzuela de Saracho	9,190
Manuel Martínez del Río	9,031
Guillermo Andrade	7,240
Total	10,491,985
Median	95,173

Note: Excludes eight titles to 818,601 ha later nullified by the Díaz government.
[a]Indicates presence of foreign capital.

leading. Companies with foreign ownership received land far out of proportion to their numbers, acquiring 3,513,138 hectares or one-third of all the land given in compensation for surveys in the six states. While they do not appear to have controlled the survey business, the firms with

Table 10 Average Compensation by State, 1883–1911

State	Average Title (ha)
Chihuahua	207,619
Sonora	117,400
Chiapas	78,548
Durango	56,877
Tabasco	29,063
Sinaloa	1,527

Source: Table 12.

foreign investors clearly were compensated disproportionately. Yet they did not receive preferential treatment from the Mexican government, because most of the land they gained title to was awarded under concessions later transferred by the Mexicans who originally held them.[37]

Finally, the level of concentration of survey company activity can be considered from two different points of view. Twenty-six firms received land for surveys they conducted in the six states, reaping 10.5 million hectares of land in compensation. For a poor, comparatively underdeveloped country that historically found it extremely difficult to mobilize capital, either foreign or domestic, these figures suggest a surprisingly broad participation in a business with a rather low success rate. But if the analysis is extended to the distribution of compensated land among the companies, a much higher level of concentration appears. As Table 9 indicates, the top three firms received nearly 6 million hectares, or 56 percent of the total; the top ten companies received 9,734,500 hectares, or 95 percent of the total. One may safely speak, therefore, of an oligopoly in the public land survey business. The following analysis of figures on a regional basis will suggest a more appropriate word—monopoly.

Concessions were usually limited to a single state or a specified area within a state. Only twelve of the fifty-nine concessions allowed surveying in more than one state, and half of the concessions (thirty) were limited to a substate zone. More were issued for Sonora—a total of fourteen, or eighteen if multistate contracts are counted—than for any other state. Forty-five of the fifty-nine concessions were issued for the northern states in the study area, because the two southern states made up only about one-seventh of the total area of the six states.

The average size of the titles granting land in compensation ranged widely across the six states, from a high of 207,619 hectares in Chihuahua to just 1,527 hectares in Sinaloa, as Table 10 reports. These variations in the size of compensatory titles (and thus of zones surveyed, since a title usually represented one-third of the total surveyed) can be attributed to the overall proportion of vacant land in each state and the degree to which it was uninterrupted by private holdings or other claims. The comparatively higher cost of surveying small zones explains why most of Sinaloa's public land was not measured until after 1900, when increasing land values made the survey more economical. Four-fifths of Sinaloa's compensatory titles were issued after 1900, while three-fourths of Chihuahua's were signed before 1887. Table 11 indicates that after 1892 the average size of compensatory titles dropped sharply while the number of titles tended to rise, another sign of the interaction between rising land values and the higher cost of surveying relatively small parcels. Table 12 shows that the surveys were distributed quite widely within each of the states; only a few districts were unaffected, and those were concentrated almost entirely in Durango.

Among those concession holders who did receive land in compensation, the most notable feature was the extraordinarily high level of regional concentration. Table 12 breaks down compensation by district for Chihuahua, Durango, Sonora, Sinaloa, and Chiapas; data for Tabasco were not available by district. Throughout the Díaz era in these states, the number of companies compensated for surveys in each district averaged only 1.57. The number of titles issued for compensation in each of these districts, excluding all Sinaloa districts, averaged just 2.3.[38] In no state did more than seven survey companies benefit, a misleadingly high figure in any event since it includes companies later acquired by others in the same state. In most districts, then, one or two concession holders dominated the survey business.

Thus not only did the most active survey companies (in terms of land received in compensation) number fewer than ten across the six-state study area, but company activity was also highly concentrated on a regional basis. While nearly every region of every state was affected by survey company activity, usually just one or two companies controlled the survey of public land in any given substate district.

SURVEY METHODS

No legislation stipulated a specific technique for surveying the public lands.[39] As a result, methods reported by the surveyors varied widely. The only controversial issue was the companies' occasional preference for

Table 11   Annual Compensation Trends in the Six States

Year	Titles	Area/Title (ha)	Total Area (ha)	Companies
1883	1	36,071	36,071	1
1884	2	671,130	1,342,260	2
1885	12	137,352	1,648,225	4
1886	14	99,232	1,389,251	4
1887	7	198,420	1,388,940	5
1888	6	21,665	129,988	5
1889	3	299,725	899,176	3
1890	15	139,486	2,092,293	6
1891	8	11,727	93,812	2
1892	2	329,829	659,657	2
1894	8	46,244	369,954	3
1895	16	7,420	118,725	2
1896	17	15,870	269,782	4
1897	1	7,340	7,340	1
1898	5	12,724	63,621	2
1900	3	1,761	5,282	2
1901	4	3,904	15,616	2
1902	9	1,882	16,941	2
1903	5	24,674	123,370	2
1904	8	57,041	456,327	3
1905	17	4,019	68,322	4
1906	23	1,688	38,833	2
1907	22	686	15,088	2
1908	3	1,410	4,230	1
1909	11	1,553	17,081	1
1910	1	2,972	2,972	1
1911	35	908	31,792	2
1912	2	2,819	5,637	1

Source: Archivo de Terrenos Nacionales, Sec. de Reforma Agraria, Mexico City.
Note: Titles later nullifed by the Díaz government are included in this tabulation.

avoiding the actual measurement of private properties, an obviously desirable approach from their point of view because it could vastly reduce their costs.

Luis Huller was the first contractor permitted to produce maps without actually measuring the private properties or their demasías. His company merely had to make a general survey of an area, subtract from the

Table 12 Survey Activity by District in Six States

District	Compensation (ha)	Companies	Titles
	Chiapas		
San Cristóbal	0	—	—
Comitán & Libertad	197,184	1	1
Soconusco	130,940	2	3
Tuxtla & Chiapa	197,184	1	1
Simojovel	0	—	—
Tonalá	162,481	1	2
Pichucalco	88,037	1	1
Chilon	120,235	1	2
Tuxtla	94,665	1	1
Palenque	0	—	—
Total	942,580	5	12[a]
	Chihuahua		
Galeana	1,022,501	2	2
Iturbide	259,742	2	2
Hidalgo	0	—	—
Bravos	393,623	1	1
Abasolo	138,764	1	1
Degollado	496,048	1	2
Guerrero	108,908	1	1
Rayón	222,943	2	2
Matamoros	0	—	—
Andrés del Río	246,311	1	1
Mina	633,528	1	2
Balleza	225,246	1	1
Allende	2,354	1	2
Jiménez	367,683	1	2
Camargo	199,284	1	2
Meoqui	334,594	1	1
Rosales	0	—	—
Victoria	0	—	—
Aldama	173,749	1	2
Ojinaga	365,204	1	1
Total	5,190,482	6	25
	Tabasco[b]		
San Juan Bautista	0	—	—
Cárdenas	—	—	—
Comalcalco	—	—	—
Jonuta	—	—	—
Cunduacán	—	—	—
Frontera	—	—	—

Table 12—Continued

District	Compensation (ha)	Companies	Titles
Huimanguillo	—	—	—
Jalapa de Mendez	0	—	—
Jalpa	0	—	—
Macuspana	—	—	—
Nacajuca	—	—	—
Tacotalpa	—	—	—
Teapa	—	—	—
Balancán	—	—	—
Total	465,002	2	16
Durango			
Durango	529,313	2	8
Nombre de Dios	0	—	—
Mezquital	373,686	2	2
Santiago Papasquiaro	347,699	1	1
Cuencamé	0	—	—
San Juan de Guad	0	—	—
El Oro	38,821	1	1
Mapimi	0	—	—
Nazas	0	—	—
San Juan del Río	4,228	1	1
Tamazula	0	—	—
Inde	107,995	1	10
San Dimas	5,119	1	1
Total	1,421,944	4	25[a]
Sonora			
Hermosillo & Ures	646,274	1	1
Guaymas	175,720	1	7
Alamos	971	1	1
Moctezuma	457,248	2	2
Altar	1,009,361	2	2
Arispe	405,845	2	3
Magdalena	161,154	2	8
Sahuaripa	195,831	2	2
Total	3,052,404	7	26

Table 12—Continued

District	Compensation (ha)	Companies	Titles
	Sinaloa		
Culiacán	42,678	4	20
Concordia	519	1	1
Mazatlán	5,182	1	3
San Ignacio	8,741	2	7
Cosalá	16,395	2	16
Rosario	0	—	—
Mocorito	29,277	3	18
Badiraguato	56,047	3	38
Sinaloa	16,712	2	17
Fuerte	62,623	5	36
Total	238,174	7	156
Grand total	11,310,586	31	260

Source: Archivo de Terrenos Nacionales, Sec. de Reforma Agraria, Mexico City.

Notes: Not all titles indicate surveys in their respective districts, because surveyors occasionally (though very infrequently) requested compensation in other districts. The total quantity of land titled in compensation is higher than that in Table 9 because eight titles to 818,601 ha later nullified by the Díaz government are included in this tabulation. Some companies are counted more than once because they received titles in more than one state. The authority for district names is Mexico, Ministerio de Fomento, Colonización, Industria y Comercio, *División municipal de la república mexicana* (Mexico: Ofc. Tip. de la Sec. de Fomento, 1889).

[a]The districts for two titles issued for land in Chiapas and Durango, granting 13,505 ha and 15,083 ha respectively, were not identified.

[b]All Tabasco titles were issued to land that included parts of one to four different districts, making an analysis by district impossible.

gross number of hectares the total privately owned hectares (by reference to the owners' titles), and submit the result as the net total of baldíos in the area. He was given permission to use this method in his 1888 survey of Sonora and Sinaloa, and it was later adopted by Gen. Francisco Olivares, the International Company, Manuel Martínez del Río, and Guillermo Andrade in their surveys of Sonora.[40] When Fomento first asked the Secretaría de Justicia for permission to grant Huller's request to proceed in this manner, it argued that "there is no [basis] for demanding that private properties be measured, nor is there a right to disturb them." Justicia apparently complied, because the same day a reply went to Huller informing him that he could submit maps without surveying the private properties, as long as he subtracted their area from the total to determine the amount of baldíos, "which is the purpose" of his contract.[41]

When a Fomento engineer inspected the work of the Mexican Land Colonization Company in 1892, he concluded that it had used the "deductive" method, because there was no evidence that individual properties had in fact been surveyed. He therefore determined that the boundary lines as given by the company were "arbitrary or of no legal value."[42] When Fomento questioned the accuracy of the firm's placement of private properties, the company's lawyer, Emilio Velasco, replied that the company was only obligated to indicate the borders of a privately owned lot as they were reported in the owner's title.[43]

In Tabasco, contractor Vicente Dardon complained that Policarpo Valenzuela's surveyors were merely asking property owners where their borders were, "and from this they have inferred the size of the baldíos." His men were not even placing markers on surveyed land, he added.[44]

Fomento tolerated the rough deductive method of surveying because the alternative would have been too slow and because this method required surveyors to accept whatever measurements property holders presented in their titles. Use of this method may have enabled the government and the companies to avoid controversies among property holders and surveyors, but it did little to serve as the check on title claims that the surveys were partly intended to accomplish. A survey of Chiapas that included the measurement of individual properties would have taken more than thirty years, estimated one Fomento official who was consistently critical of the companies' work.[45] Even driving stakes to mark corners was often impractical, as was traversing the often difficult terrain. In 1912 a government surveyor in Tabasco reported that even in the dry season, his men could not walk more than fifty meters from the banks of the rivers without sinking up to their waists in mud. Timber had to be cut so trails could be blazed; it might take two or three days to get from one *montería* (timber stand) to another, "and many times eight just to find a rancho."[46]

Thus the companies seemed more likely to check property limits in Chihuahua and Durango, and in Sinaloa after 1900, than in the southern states sampled in this study, because of the far more challenging natural barriers to a field survey that the southern states presented. Antonio Asúnsolo's company reported that in its survey of Durango, for example, the lack of opposition resulted primarily from the company's practice of sending surveyors to each private property before starting the survey of the baldío to examine the lot and "obtener el terreno baldío indisputable" by surveying the land that was obviously private. Any adjacent land whose status was uncertain would be segregated so that a court could determine its ownership.[47]

In some surveys so many titles were questionable that the baldío of a zone was not computed until the authorities ruled on the validity of the titles. When José L. Galán was commissioned to resurvey Manuel Peniche's surveys of the Arizpe and Moctezuma districts in Sonora, he appended a list of putative property owners, pointing out that some produced no titles at all, others presented titles that clearly were not legal, and that some titles were merely "doubtful." As a result, "the Company cannot compute the land that is really baldío, until the Supreme Government achieves a settlement regarding all these titles."[48]

At times, an incomplete or inaccurate survey was preferred if the alternative meant provoking a hostile reaction from the inhabitants. Policarpo Valenzuela's company found 8,462 hectares of baldío in Tabasco's partidos of Cárdenas and Cunduacán, but nearly three-fourths of it was a demasía that the surveyors feared detaching from some properly titled land. As a Fomento functionary explained, the engineers decided to leave it alone "because of the difficulty that this delicate operation presented for the surveyor, [the land] being occupied by a great number of individuals, who no one certainly wants to disturb." Consequently the demasía of 6,128 hectares was deducted from the baldío, leaving just 2,333 hectares for the government and the company to divide.[49]

ECONOMIC FACTORS

The companies' technical and scientific mission created potential conflicts of interest not only with the commercial character of the enterprise but also with the need to interpret the legal documents that were presented by property holders, a task that was often avoided by the companies. The chief surveyor of Jesús Valenzuela's company refused to rule on the legality of documents presented at San Borjas in 1884. The survey party not being a court that could rule on the documents' validity, "but a scientific and topographic commission, to record by means of geodesic constructions, the position and surface of the properties, their duty was circumscribed in legal matters to express who possessed [the properties] and with what documents they possessed them."[50] Similarly, the surveyors for Manuel Peniche pointed out that even though they believed many of the properties indicated on the map they produced were in reality baldíos, they were letting the government determine the validity of the claims—a point that would later be seconded by Galán, as indicated earlier. When the government urged the Mexican Land Colonization Company to arrange composiciones in Sonora, the company's lawyer, Emilio Velasco, declined on the grounds that the government was better suited to handle "an extremely delicate question that it is as well not to touch except with great prudence and without speculative interests."[51]

How closely companies examined legal documents, whether they chose to survey private properties, and which particular survey method they adopted were all choices that were ruled by the same criterion applied to the choice of land to survey—cost. To the chief surveyor of the Mexican Land Colonization Company operating in Chiapas in 1891, the profit-making character of the operation was paramount. Thus the choice of survey method was dictated by its cost. Speed, accuracy, and economy were all relevant factors to consider in deciding the method. Keeping costs down was particularly important, he added, because "the work was of a commercial character rather than scientific." Cost cutting also provided an incentive to the companies to avoid needlessly provoking property holders. The Mexican Land Colonization Company's surveyors were instructed to treat property owners "with kindness and courtesy," to make them understand that "no one was trying to dispossess them," and to respect "the local authorities and to take great care not to have quarrels with the Indians."[52]

To lower costs further, the companies sometimes avoided surveying relatively small areas of public land, such as those surrounded by numerous property owners who may have had conflicting claims to the baldío. This practice was criticized by Encarnación Dávila, who received a contract for Coahuila in 1885 and proposed to undertake a detailed cadastral survey of Coahuila and Nuevo León. The government, he wrote, would have the advantage of getting much public land "that is mingled with the corrupted boundaries of the private property owners, and many times through malice and bad faith on the part of the proprietors." A real cadastral survey would achieve in fact what the government was vainly attempting to achieve in theory with the survey companies. The latter, "seeking only profit, do not survey small properties, nor the bit of left-over land that would not cover expenses, and consequently they leave it without surveying it." As a result the possessors not only keep what should be the government's, but the government does not even get to ascertain who owns what land.[53]

The engineers hired to carry out the surveys were sometimes partners of the companies and were paid in land, just as the companies were. Grounds for a conflict of interest thus emerged, with the engineers' duty to produce an accurate survey possibly conflicting with their desire to enrich the company or themselves. For example, William P. Morrison, a surveyor hired by Ignacio Sandoval's company in Chihuahua, was a partner in the company and part owner of 620,154 hectares it received from the government in compensation. When an official of the revolutionary government reviewed the survey in 1919, he used Morrison's status to argue the unreliability of the survey.[54]

Other companies failed to compensate their surveyors enough to keep them. Federico Nieto quit surveying for Bulnes Hermanos in Tabasco because of the low pay.[55] When another Tabasco company, Policarpo Valenzuela's, complained that it was unable to hire enough surveyors because the federal judge was assigning them to other work, the judge retorted with a reminder of the law of supply and demand. The company would not lack surveyors if only it paid them more.[56]

Little evidence bearing on the companies' profitability could be found. As one would expect, the owners never lost an opportunity to complain to Fomento that they were losing money, while none ever admitted to prospering. Jesús Valenzuela's company took two years to survey Chihuahua's Cantón Iturbide; the judicial record alone totaled 281 pages, and the technical reports came to 77 pages. José Valenzuela, the company's agent, complained of "much work and substantial costs . . . instead of producing for the company, it has been a real and true loss."[57] The cost of surveying was so high in Sinaloa in 1906 that the Sinaloa Land Company preferred to arrange composiciones with landholders whenever possible. One of the company's agents wrote from Culiacán to partner Luis Martínez de Castro that whenever it was necessary to resort to a survey and attendant court action, the company had to locate the titled portion of land and their demasías, negotiate with Fomento on the division of the baldíos, pay engineers and other workers, and spend time getting landholders to agree to the survey. All of this meant that "profits come down to zero for the company, which is obliged to keep properties that in this State are not worth anything, nor is there anyone who would buy them." The company also faced the additional costs of unnecessary litigation provoked by crooked lawyers who played on the credibility of landholders, then secretly sought money from the company to withdraw the opposition. The firm had refused all such offers, the agent said.[58]

CONCLUSIONS

The Mexican government granted survey concessions to virtually anyone who applied; political, financial, and technical credentials were evidently irrelevant. But a swift shakeout by means of concession sales and numerous apparent business failures led to control of the survey business by a few major firms. Just ten companies, including some with a significant but indeterminate level of participation by foreigners (British or American investors), practically controlled the public land survey business in the six states. Some of those who dropped out did so by selling their concessions to others; perhaps this was the reason they acquired the concessions in the first place, although the record does not speak to us on this point. Others clearly tried and failed—they ran out of cash or were unable to handle the technical demands. In those cases the companies, unable to overcome the enormous risks and meet the high costs of surveying, either quit or had their concessions invalidated for faulty performances. Many of these risks—including litigation, bureaucratic obstacles and delays, attacks by hostile Indians in the North and the almost insuperable natural obstacles encountered in the jungle terrain of the two southern states—are explored in detail in chapter 3.

As the company profiles in Appendix A indicate, some of the companies that survived were headed by individuals who were exceptionally close to the levers of political power, but these connections were not necessarily a condition of success. Others appear to have had little sway in the capital; they probably enjoyed access to both financial and technical resources that were beyond the means of most of the firms. At this point we can do no more than speculate about the relative significance of these and other relevant variables: inherited wealth, kinship ties, contact with foreign investors, and access to other people's capital, among others.

Yet some conclusions do emerge. First, a government concession alone was not enough to make a man rich. More than fifty enterprises received concessions either directly from Fomento or from an original concessionaire. Nearly half the concessionaires failed to carry out a single successful survey. Many who did never approached the enormous holdings that the top surveyors amassed. Political connections, a theme that runs throughout the biographies in appendix A, were obviously important but may not have been essential. They were certainly not sufficient, as the failure of the Compañía Descubridora with its cast of politicians—including Porfirio Díaz and Carlos Pacheco—generals, and their flunkies proves, and as the bankruptcy of the International Company also indicates. (See their profiles in Appendix A.) Furthermore, even the most successful concessionaires were at times treated with varying measures of hostility and contempt by governmental authorities, and were often disciplined for flouting the law, as will be further demonstrated in chapter 3.

Second, the financial data assembled in the profiles in Appendix A also disclose a tendency for nominally independent entrepreneurs to share resources. In some cases the extent of financial interlocks is extraordinary. One group of investors more or less controlled the surveys of Chihuahua and Durango (and perhaps of other northern states that were not part of this study). The group was composed mainly of Antonio Asúnsolo, Tomás Macmanus, Plutarco Ornelas, Telésforo García, Ignacio Gómez del Campo, Ramón Guerrero, and (by way of the Meza Hermanos concession) Porfirio Díaz himself, along with ex–Fomento chief

Carlos Pacheco and a legion of generals. The concession held by Jesús Valenzuela, which alone collected nearly one-third of all the land awarded in compensation in the six states, could also be associated with this group, since Telésforo García was one of Valenzuela's principal partners. The connection could be extended still further if one counts the concession awarded to Valenzuela's sister Natalia in Sinaloa. Every one of the concessions (except for the latter, which was acquired from the original concessionaire) was awarded during the González administration.

The Chihuahua-Durango group is noteworthy for the *apparent* (one cannot logically eliminate the possibility of foreign participation in any company) absence of foreign capital, since Chihuahua, according to Wasserman, was inundated with U.S. investment capital when the surveys were taking place in the 1880s. That the survey business in Chihuahua remained a domestic affair in a state that had been the target of more foreign investors than any other, except possibly for Sonora, was a sign of the vitality of its economy.[59]

Overall, foreign capital played a significant role in the surveys. American and British investors joined in the survey business by either buying concessions from Mexicans or acquiring shares in concessions that had been awarded and retained by Mexicans. Thus their participation does not seem to have been the result of government preference but of normal market activity. The level of foreign participation in the companies ranged from nearly 100 to 0 percent; while only six of the twenty-six companies that received land in compensation had foreign investors, they accounted for one-third of all land granted in compensation. Foreign capital was no guarantee of success, however—both Huller and the International Company were unable to complete their obligations and were forced to sell out.

Finally, a marked tendency toward geographic exclusivity developed everywhere in the six states, with one or two companies dominating the surveys of substate jurisdictions, despite the government's early efforts to issue multiple overlapping concessions. The gradual narrowing of the field was a result of the play of the microeconomic factors (access to capital, technical resources, etc.) mentioned earlier.

The state's evolution to the full-blown interventionism of the twentieth century was in midpassage during the late nineteenth century. The governments of Díaz and González made adroit use of the concession tool to balance competing interests while achieving the rapid survey of the public lands. If capitalist influence on the government had reached its apogee in the Porfiriato, as Camp hypothesized, it was an influence that was tempered and directed by the government itself. The high value that the Díaz government placed on rural peace induced it to create an elaborate system of judicial and political controls on the surveys. The authorities were well aware of the surveys' potential for fomenting unrest, and both the government and the companies had strong incentives to avoid antagonizing property holders. But the accuracy of the surveys suffered as the companies—with Fomento's occasional indulgence—responded to two interrelated conditions. The first was the desire to avoid unnecessary conflict with property holders. The evidence presented in this chapter shows that far from riding roughshod over property rights, the companies (and probably the judges as well) often accepted even questionable claims and sometimes used technically crude methods to steer clear of costly disputes and litigation. Second, high-quality surveys were expensive. Since the contractors were not compensated according to the costs they incurred, they had no incentive to exert more than the minimum effort necessary to establish the limits of the baldíos they discovered. These incentives were shared by Fomento to a considerable extent, given its interest in swiftly establishing the limits of public land to raise revenue quickly, while avoiding as much as possible the inevitable conflicts. (Yet as we shall see, the government generally supervised the companies quite closely, requiring them to observe not only the procedural formalities of the surveys but also the legitimate rights of property claimants.) An evaluation of the overall quality and fairness of the surveys, taking into account both the archival record and secondary sources, is presented in chapter 6.

This chapter has shown that during the Porfiriato both the private and public sectors were still finding their way; neither was dominant. The "modern economic actors" that Guerra correctly points to as among the creations of the state benefited (in the case of the survey companies) only to the extent of receiving a government concession. The subsidies and favoritism that would characterize the postrevolutionary state were not altogether absent during the Porfiriato but they were much less evident than they would become. The survey companies survived, failed, or prospered largely according to conditions outside the public sector. Subsequent to the fragmentation and rapacity of state-business relations of pre-Porfirian Mexico, a genuinely capitalist class was beginning to develop whose growth would be abruptly stunted in the aftermath of a revolution that elevated the state to a pinnacle undreamt of by Porfirio Díaz.

3 State Management of the Surveys

The Clash of Public and Private Interests

 This chapter explores in detail the relationship between the Porfirian state and one segment of the country's entrepreneurial elite—the survey companies. As Mexico entered fully into its transition to capitalism, how much autonomy did the state exercise in its relations with that elite?

There can be little doubt that the state ruled on behalf of the generally prosperous men who were willing to risk investing in survey concessions.[1] Such investors were among the elite whose interests the state sought to encourage and defend in every possible way, because their wealth, talent, and propensity to take risks were considered the keys to Mexico's economic development. Nevertheless, the regimes of both Díaz and González clearly saw themselves not as the mere administrators but as the directors of the survey companies' activities. Both regimes took action that contradicted the companies' private interests when other issues, such as "la paz social," were considered paramount.[2]

The political environment in which both governments operated was scarcely uniform over the three and one-half decades during which they held office, and their decisions were therefore subject to political considerations that varied over time. For example, most of the survey concessions were awarded before 1888, the year of Porfirio Díaz's third inauguration. But it was not until that year (the beginning of what Cosío Villegas called "El Necesariato") that Díaz was able to fully consolidate his power in the presidency, a full twelve years after the coup by which he had attained power in 1876. For four of those years the post had been held by Manuel González, who if he maintained the political organization begun by Díaz in the latter's first administration, nevertheless ruled the country as his own man, and not as Díaz's. The sharp polarization between *porfiristas* and *gonzalistas* that characterized politics in 1884–85 is further evidence of the dispersion of political authority in that decade.[3] About half of all the land given in compensation (5.4 million hectares) was titled under concessions that had been awarded before Díaz had even begun his second term of office.

Just as the state's behavior was subject to a variety of interests and pressures, the very self-interest that distinguishes private enterprise encouraged a diversity of responses from the surveyors that are not susceptible to facile generalizations. As profit seekers, the companies sought to cut costs whenever possible. When the state, a landholder, a community of peasants, or a competing company threatened to raise those costs by means of further litigation or even violence, the contractor had to decide whether to pay the costs; he often chose not to. The companies' behavior was also constrained by the variety of often-conflicting interests involved in the surveys, as represented by local authorities, the federal government, other surveyors, entrepreneurs of all kinds, landholders, and traditional communities.

The overall legal and constitutional framework governing the disposition of the public lands has already been described in chapter 2. Here we will analyze the actual dynamic of state-company relations in the following key areas: (1) the level of the government's control of the survey companies, particularly enforcement of its own regulations; (2) the companies' relations with local political authorities and the nature of the contact between federal officials and local governments in questions involving the companies; and (3) the federal government's role in the rivalries and conflicts that emerged among the survey companies themselves. One issue, the usurpation of privately held land by the survey companies, merits special treatment and will therefore be discussed in the next chapter.[4]

GOVERNMENT ENFORCEMENT OF SURVEY CONTRACTS

When the companies submitted their surveys to Fomento for approval, the ministry inspected them for evidence of juridical or technical aberrations ranging from relatively minor oversights to the misplacement of property lines. Penalties, including the abrogation of the contract or simply the denial of compensation, were frequently assessed as a result. In the following pages we will see how Fomento responded to the most common types of irregularities, evidence that will be used to assess the relationship between the companies and the state—particularly the state's freedom to enforce legal and contractual obligations in its dealings with the private sector.

Fomento routinely withheld its approval of surveys for which the documentation was incomplete, even delaying action until it received such relatively minor data as the required physical description of the surveyed zone, or the surveyor's classification (as first, second, or third class) of the land.[5] Scores of surveys were sent back for revision when Fomento staff found defects, omissions, or contradictions in documentation. For example, Jesús Valenzuela's survey of the municipalidad of Moris in Cantón Rayón was returned in 1885 with this sarcastic observation: "they are not in conformity with the law in various ways, there being so many shortcomings in the proceedings and in the map, that one could well say that they are not surveyed."[6]

Another survey was returned to the federal court in Durango when Rafael García Martínez inadvertently gave the pueblo of Bocas four square leagues for its fundo legal instead of the one to which it was entitled.[7] In another case, when William T. Robertson's representative, Manuel Martínez del Río, asked Fomento for compensation for a composición in 1892, the ministry pointed out that some of the property-line descriptions had not been signed by the adjoining property owners, other adjoining owners were not summoned, the titles seemed to cover only about half the land their owners claimed, and the survey report was too vague. In a separate survey involving Robertson, Fomento requested a copy of the surveyor's field notes when his report indicated that he had measured thirty-nine kilometers in one day, which sounded suspiciously high.[8]

These few examples could be multiplied dozens of times throughout the period. Yet surveys in Sinaloa and Sonora seemed to be particularly vulnerable to this kind of careful scrutiny in Mexico City, for three reasons. First, surveys in these states, but especially those in Sinaloa, tended to involve smaller lots and to be much more numerous than surveys elsewhere, so there were more opportunities for the government to catch errors. Second, the Sinaloan surveys took place late in the Porfirian period, when there were many more regulations with which to contend. Finally, the supply of unsurveyed public land was diminishing rapidly, so the government's return on the survey company business (i.e., the two-thirds it got to sell) declined absolutely after the mid-1890s, thus giving Fomento less incentive to overlook defects.[9]

Those surveys conducted after 1900 by Luis Martínez de Castro and his successor, the Sinaloa Land Company, were the most closely studied of all by Fomento; files were returned to the federal court to rectify the slightest oversights. In some such cases, the ministry rejected surveys because angles were drawn too sharply, or because the surveyor failed to specify the values of the angles, or because of the quality of paper used for a map—objections that had never before been raised.[10] Many, perhaps most, of Fomento's objections to the two companies' work in Sinaloa, however, were related to the sufficiency of evidence that a particular adjoining property holder had approved the placement of the baldío surveyed. Technical errors by the surveyors and noncompliance with procedural formalities constituted the grounds for most of the other objections.

A common form of discipline against the companies was to declare survey contracts or authorizations null if the contractors failed to begin within the time required by the contract. When Fomento learned in 1890 that Rafael García Martínez had still not begun work in Sinaloa under two authorizations granted in 1883 and 1884, it informed him of the president's declaration that both contracts were "lapsed and baseless."[11] The failure of Telésforo García and Jesús Meraz to present the "documents and maps that justified their work" prompted Fomento to nullify their contract in March 1887.[12]

Antonio Asúnsolo's authorization to survey in Durango was lifted because it was late in starting, a delay that hurt the "interests of the Government, because it needs those lots and this prevents it from carrying out transactions involving them." The company claimed that it had begun the survey on time and that the federal judge had erred in ruling it had not done so. Like most government decisions, Fomento's was negotiable, particularly since Manuel Romero Rubio, the secretario of gobernación and the president's father-in-law, was associated with Asúnsolo's Durango operations (see Appendix A and chapter 2). General Pacheco informed his fellow cabinet minister, Romero Rubio, of the decision "for your information, and so that if you consider it appropriate, you may interpose the resources that you think conducive to the salvation of that Company's interests." Asúnsolo received a similar note, and in April 1886 both men were invited to General Pacheco's home at 5 Humboldt Avenue for a 9 A.M. meeting "to confer about the matter." Since the firm later received title to land in Durango for its survey there, the men apparently worked out an "arreglo."[13]

When a company erroneously included titled land in a survey and did not discover it had done so until after it had already received title to the same land (as its one-third in compensation), it usually asked Fomento for compensation elsewhere as a replacement. But the government habitually rejected such requests, on the grounds that the company's job was to find baldío—an inaccurate survey did not deserve compensation. A typical response to such a request was that of Gen. Carlos Pacheco, the secretario of Fomento, to Gómez del Campo in 1887: "the Government's obligation ends from the moment when the surveys you made, which were understood to be good enough on which to base various transactions, have turned out to have important errors."[14]

Sometimes Fomento had so little faith in one of its contractors' surveys that it hired its own engineer to resurvey the land under a so-called rectification contract. Such was the case with Asúnsolo's survey of the partido of Santiago Papasquiaro in Durango. The engineer hired to resurvey the land, Manuel Tinoco, filed his report in September 1893, about a year after he was hired. Tinoco found "a notable concordance" between his survey and Asúnsolo's except in the area designated as zone two, where the company committed "absent-mindedly, large errors." With that exception, Tinoco concluded, the company's measures "are acceptable, and even exact." Asúnsolo, he added, scrupulously respected property holders' acquired rights to land and carefully summoned all known owners to present their titles.[15]

Another rectification was undertaken after Manuel Peniche surveyed parts of Sonora, and this one was done at his own expense. Peniche was forced to return his title to 301,934 hectares he received in compensation for the defective survey, which Fomento determined included land already surveyed by other companies.[16] The ministry informed him in May 1887 that, "having sound reasons to believe that all of the surveys that that Company [de Tierras de Sonora] has carried out in the state of Sonora suffer from many irregularities, the President of the Republic has had to send a Government engineer to rectify them at the Company's own expense."[17]

The government hired two U.S. and one Mexican civil engineers, William J. Glenn, Henry O. Flipper, and José L. Galán, to correct Peniche's survey of Arizpe and Moctezuma districts. They filed a detailed report in October, noting the omission of "varias propiedades" from Peniche's maps. The biggest error Galán found was in Peniche's line joining the sierras of Pinitos and La Púrica, which was off 1 meter for every 262 meters; the smallest was 1 for every 11,903 meters, with an average error of 1 for 946. Galán did not find this excessive: "Although strictly speaking these errors would not be scientifically tolerable, I think they may be overlooked, and work of this nature that has given rise to relatively small errors can even be considered quite good." Of the properties listed for Arizpe, many were not titled at all, or their titles were of doubtful legality, which to Galán suggested that a measure of leniency was in order in evaluating Peniche's work, since "the Company cannot compute the land that is truly baldío until the Supreme Government resolves the question of all those titles, as well as land that had been baldío and was sold after the first measurements were carried out."[18] After trimming 222,827 hectares off Peniche's original claim to have surveyed 905,802 hectares in Arizpe, he was compensated with 227,658 hectares (one-third of the newly calculated area of 682,975 hectares), and received an additional 305,877 hectares for Moctezuma.[19]

When Fomento found "graves irregularidades" in Policarpo Valenzuela's surveys of the districts of Jonuta, Montecristo, Balancán, and Tenosique, it ordered a new survey in 1891, to be carried out by the Comisión de Límites Entre México y Guatemala.[20] The results were shocking. Rivers were out of place and borders did not coincide in a single point. (The Guatemalan border was 8,900 meters too far to the east—"it seems that 55,000 hectares were taken from Guatemala"—and 250,000 hectares were "taken" from Chiapas.) Some landmarks were off by as much as twenty kilometers.[21] Valenzuela agreed to resurvey the land, but when he still had not done so by 1900, the federal government filed suit demanding the nullification of his title to 247,777 hectares in Jonuta and Balancán, which had been issued in compensation in 1890. Fomento charged that the company had been titled considerably more than the one-third it was due. Valenzuela's defense was that he was unable to determine the border between Chiapas and Tabasco, that the errors were the engineer's fault, and that he had already begun to rectify the survey. The court ruled in favor of the government, nullifying the title and assessing a $12,831 fine, which the contractor paid in 1903.[22]

It was not uncommon for Fomento to simply reject a defective survey, in which case the company had no means to recover its investment, since no land was given in compensation. Perhaps the unluckiest contractor of all was Gen. Hipólito Charles, who was authorized to survey in Sonora in 1889. Of the twenty-seven lots he surveyed during the next eight years, twenty-three were rejected by Fomento—because they had already been surveyed by Manuel Peniche! The luckless general surveyed more than 180,470 hectares, of which 149,558 (about eight out of every ten) were rejected. He ended up with titles to just three lots totaling 30,852 hectares.[23]

Of all Fomento's enforcement actions, none was more time-consuming and demanding than that required by the surveys of Chiapas, which were probably the least reliable of any conducted in Mexico. Unlike the other states in this study, no more than one survey contractor was ever authorized to work in Chiapas at any one time, but the contract itself changed hands four times. Let in 1886 to Andrés Gutt, who transferred it three days later to Luis Huller, the contract was then passed on to the International Company, the Mexican Land Colonization Company, and the Land Company of Chiapas. It was subjected to numerous revisions along the way and was also tied to a colonization contract awarded to Huller.

That the survey of Chiapas should have been so vulnerable to criticism was probably due in part to the nature of the climate and the terrain.

Even the most conscientious surveyors would have been challenged by the high heat and humidity, rugged mountains, and impassable rain forests, not to mention the epidemics of measles, scarlet fever, whooping cough, typhus, cholera, and smallpox—the last three common throughout the state in the 1890s.[24] There was also the matter of what one Fomento official called the "absolute lack of roads."[25] José Covarrubias, the head of the Fomento section that handled surveys, and the companies' harshest critic at the ministry, observed in 1895 that because of the natural obstacles presented by Chiapas, a real survey of just six of the state's eleven departments would take thirty years. Simply mapping the private properties in the six departments would take four to five years, he estimated, and even that would not represent "a survey according to the true definition of the word," serving only as a guide for locating baldíos.[26] Chiapas was probably the best argument for a government-sponsored and funded survey, since no private firm could be expected to invest the resources that would have been required to produce what Covarrubias considered a real survey of the state.

Starting in the early 1890s and continuing until after the overthrow of President Díaz, Fomento would find much to criticize in the Chiapas companies' work. After the Mexican Land Colonization Company surveyed parts of the departments of Comitán and La Libertad in 1890–91, the local prosecutor informed the federal judge of several errors and omissions, which apparently led to a close review of all the company's surveys (their approval was still pending) in the departments of Tuxtla, Chiapa, Pichucalco, Comitán, and La Libertad by a Fomento engineer, Enrique Hijar y Haro. The field operations, Hijar y Haro found, were so substandard that they could not be called surveys, though he pointed out that the company was only following a practice already approved for Sonora and Sinaloa (though not yet for Chiapas)—simply to draw a map of private properties without actually measuring them, then deducting the amount of land specified in the titles to arrive at the quantity of baldío.[27] The company was also criticized for respecting all private property claims, even those for which no title was produced. This was a novel complaint, but the practice irritated Fomento because it defeated the purpose of the surveys, namely the discovery of untitled land, although the company probably found it a speedier and less costly alternative to litigation. In any case, Hijar y Haro found that the private property limits accepted by the company were "arbitrary and of no legal value," meaning that the lots requested by the company in compensation were also badly defined.[28]

However, since the company was only following a practice permitted it in Sinaloa and Sonora, Fomento eventually decided to accept the surveys of Tuxtla, Chiapa, Pichucalco, Comitán, and La Libertad in February 1893.[29] By November, however, the titles had still not been issued, and a second report was prepared, pointing out some of the same defects discovered by Hijar y Haro, and calling attention to new ones. Private property limits were "false, or at least inexact," which would present "almost insuperable" problems when the government tried to sell public land. Worse, the company had not questioned numerous public land claims that clouded the status of the baldío. The list of private properties contained so many omissions that "they lead one to think that the surface claimed as baldío is illusory." In Comitán alone, eleven fincas were listed without their areas being given because the owners failed to present titles, even though they were required to do so by the law of July 22, 1863. More than thirty other fincas were found by Fomento that were not even mentioned by the company. The reviewer concluded that the survey revealed neither the location nor the extent of baldíos, leaving Mexico "as backward as before. . . . [The] operations are, in a word, useless."[30]

But no Fomento functionary could match the mastery of detail, knowledge of law, and logical powers of Emilio Velasco, the company's lawyer, easily the most impressive mind of any encountered in this study. Retained first by the International Company and then by its successor, the Mexican Land Colonization Company, Velasco probably saved both firms from losing their concessions through his ability to argue his way out of trouble. Some properties were not identified because the owners failed to produce titles, Velasco explained, urging Fomento to hire its own engineer at company expense to locate them, and give him the authority to force owners to cooperate. Further, the company's contract required it to do no more than indicate the borders of property on a map—it had no authority to identify demasías. As for the pending public land claims, all the surveyed land had already been claimed by the company, which was rightly ignoring those claims that were filed and then abandoned.[31]

At the same time, Velasco wrote to Chiapas Gov. Emilio Rabasa to oppose state legislation requiring the return of property unjustly transferred to the survey company. Velasco complained that local authorities failed to help the company overcome the "bad attitude of some property holders who avoid producing their titles" and offered to return any unjustly acquired properties immediately. He divided the claims against the company into three classes: (1) those by property holders who usurped land and then tried to get the surveyors to accept the claim, refusing to produce titles because none existed (such as the case of one General Escobar); (2) public land claimants whose claims were filed after the company's; and (3) public land claims done before the company's autho-

rization to survey that are "fraudulent and of bad faith." Velasco challenged the governor to find one owner whose property was not marked
on a map, and said the company had to confront "greed" and "evil pas-
sions inspired by the desire to take over baldíos," including the attempted
murder of a company agent who was badly wounded. If the state really
wanted a good survey, it should require its local judges to force property
owners to produce their titles.[32]

In late February 1894 Fomento accepted Velasco's proposal, naming
an engineer to find, at company expense, properties not mapped, while
directing the federal judge and prosecutor in Chiapas to see that landowners presented their titles and asking Governor Rabasa to ensure the
cooperation of local authorities. Alberto Amador, a civil engineer, accepted an appointment to rectify the survey in April. Velasco, while ordering company agents to cooperate with Amador, reminded Fomento
that unless the government took effective steps to force landowners to
present their titles, Amador's mission would fail completely. Individuals
who refused to show the proper documentation probably stole the land
they were on and did not want to be discovered, and thus had "an interest
in the failure of the surveys." Governor Rabasa ordered the state's jefes
políticos to require owners to cooperate with Amador by presenting their
land titles to him.[33]

Amador filed his report in September. In the department of Soconusco, he found that the company had omitted thirty-three properties totaling 26,974 hectares; some of those the company did map were off by
as much as six kilometers. Yet to the fifty-four properties identified by
both the company and Amador, the former attributed 63,075 hectares
and the latter 48,104 hectares.[34] José Covarrubias, the Fomento official,
angrily observed that at a cost to the company of just $5,000, Amador's
work in Soconusco was more complete than the survey done by the company, which received 34,957 hectares for its survey of Soconusco, land
that Covarrubias estimated was worth more than $100,000.[35] But ac-
cording to Velasco, Amador's "survey" was conducted entirely from his
desk, where he simply transferred data from titles to maps, while the
company's survey was based on fieldwork and was therefore more reliable. In any case, the company was apparently penalized for mistakes it
was alleged to have committed, since the title to 34,957 hectares was nullified and replaced in 1897 by one for 30,955.[36]

Amador resigned before rectifying any more surveys, and a new agreement was reached between the government and the Mexican Land Colonization Company in January 1896. The company would resurvey all
the surveyed land remaining at the disposition of the government[37] in the
departments of Soconusco, Tonalá, Pichucalco, Tuxtla, Chiapa, and most

of Comitán. But the company would have to accept any reduction in its share if the survey was successfully contested by any property holder, without having the right to replace any of it from the government's share. By 1903 the company had only resurveyed in Soconusco and Pichucalco, having been titled 202,121 hectares in compensation and another 190,024 that it purchased.[38]

When Fomento sent its own engineer, Ismael Loya, to Chiapas at the end of 1901 to verify that the government's share was indeed not owned by anyone, Emilio Velasco reminded the ministry that the company had allowed the government to choose any land it wanted for itself, even though it was not obligated to do so, and that the government "always picks the best." Moreover, the company would not assume responsibility for people who were squatting on baldío, a common occurrence in Chiapas.[39]

Loya telegrammed Fomento in May 1902 that, after observing much of the land that he was to take possession of in the name of the government, "in its totality it is made up of land that is completely inaccessible, hills and extremely rugged elevated peaks . . . , this being the worst land in this department." Fomento told him not to worry about the quality of the land, since the government had picked it; his job was to verify its limits. Loya found one finca totaling 8,701 hectares; Velasco said the title had never been presented. A difference between Loya's and the company's survey of more than one-half of a degree in direction was found. They also differed slightly in the area assigned to one lot. In his defense, the company's surveyor pointed out that "it is well known" that the same surveyor can come up with two different figures upon resurveying property, something that is even more likely when two different men perform the operation. He also pointed out that markers placed by the company could well have been lost in the region's abundant vegetation.[40]

Associated with the survey was the colonization contract that had been awarded to Huller. It was also passed along to the successor companies and was replaced by a new colonization contract in 1897 and again in 1913 during the government of Gen. Victoriano Huerta. Although the scope of this research project excludes any detailed treatment of colonization, it is worth noting that none of the companies established the number of colonists required under the contracts, a source of continual tension between the companies and Fomento. The ministry appeared to be no less concerned about enforcing the colonization contracts than the survey obligations. The 1897 contract required the company to settle 848 colonists (3 out of 4 had to be non-Mexicans) within ten years, and when the deadline had passed with no more than 47 in place, Fomento told the company it owed a fine of $80,100 (as specified in the contract) and would lose its $5,000 performance bond. The company, claiming it had actually settled 345 colonists, acknowledged that it had found the task "imposible." During the Huerta administration a settlement was reached under which the company, instead of paying a fine, agreed to return to the government "for the diverse pueblos of the state" twenty-eight lots totaling 40,443 hectares. That deal was nullified by the regime of Gen. Venustiano Carranza, ironically (given the supposedly "counterrevolutionary" character of General Huerta and Carranza's status as "first chief of the revolution") putting General Huerta's donation of land to the pueblos in jeopardy; the final settlement will be discussed in chapter 6.[41]

Fomento also took action to punish Antonio Asúnsolo's company for failing to meet its colonization obligations. In 1890 the government agreed to sell Asúnsolo 430,000 hectares of terreno nacional in the Chihuahua cantónes of Balleza and Jiménez for twenty centavos a hectare on the condition that colonists be established on the land. Fifteen years later Fomento discovered that the company, which had never actually been titled the land, still had neither paid for nor settled any colonists on it. Yet the company had been renting out the land and otherwise exploiting it. The firm's appeal to President Díaz in 1905 that it be allowed to fulfill the terms of the contract and keep the land was rejected. The government insisted on collecting the full fine of $17,200 (as specified in the company's colonization contract), canceling the contract, and liquidating the $1,500 performance bond the company had on deposit. The company was unable to pay the fine, but the government collected the amount due by auctioning land owned by the firm in the former cantón of Victoria in 1906.[42]

THE COMPANIES' RELATIONS WITH LOCAL AUTHORITIES

Once authorized by the federal government to survey baldíos, the companies also had to receive a separate authorization from the federal district judge in the particular state to survey specific zones. The survey was supervised by the federal court, which in turn often directed local magistrates to provide a judicial record of the day-to-day fieldwork of the survey. This record, which included such procedural matters as the summonses to property holders to present titles and any protests filed against the survey, was then forwarded to the federal judge.[43] After the publication of notices giving additional opportunities for protests and the completion of other formalities, the judge sent the entire file to Fomento. Control over surveys was thus fairly decentralized, with federal judges having a great deal of power if they chose to exercise it. The survey itself remained largely a federal responsibility. State governors and such local authorities as jefes políticos appeared to have scarcely any direct, formal influence on the operation of the survey, though the governors were routinely informed of all contract awards, suspensions, and nullifications. The exception was Sinaloa, where (as noted earlier) the governor was frequently consulted by Fomento about the occupation of surveyed public land.

Given the important supervisory role of the federal courts, it is not surprising that the local authorities with whom the companies conflicted most frequently were the federal judges. The federal judiciary, like the legislative branch, was (though not in constitutional terms) a dependency of the executive.[44] That conflicts could arise at all between the companies and the courts, therefore, is consistent with our finding in this chapter and in chapter 4 that the federal executive often clashed with, and disciplined, the individuals upon whom it had bestowed the privilege of a survey contract. Nevertheless, what stands out above all is that conflicts of any kind between the companies and local officials, including judges, were quite rare. Those that occurred were usually over the judge's admission of public land claims on land designated by the companies for survey, a constant source of contention that Fomento only occasionally seemed to take seriously.[45]

During the survey of the Chihuahua municipalities of Nonoava and Sisoguichic by Gómez del Campo and Guerrero, the contractors complained to Fomento in April 1883 about the hostility of the federal judge and his continued admission of public land claims in the company's survey zone. The firm feared the belligerent character of the Chihuahuans, individuals easily provoked by the judge, whose behavior would have "unfortunate consequences, because the ignorant peasants who take out a requisition to survey baldíos at the same time as the [survey] company, will take up arms against the company's survey parties. We do not exaggerate, we know the people of the State and this is what we have told the Judge who does not know anyone here." By forcing the company to file a separate action to counter each public land claim, the judge was holding up the survey. In this case Fomento told the judge to stop admitting claims on land in the area already designated for a survey, since doing so would make it impossible to survey the land. The federal government, Fomento reminded the judge, would appreciate expeditious proceedings, "it being necessary that the Government have at its disposal as soon as possible the two-thirds that correspond to it in the survey."[46]

Jesús Valenzuela's company complained twice of lack of cooperation by a federal judge, who was allowing opposition to the survey based on questionable claims (including one by Luis Terrazas, to whom the judge was "dócil"). Valenzuela wrote that "the difficulties that we have run up against in our work have been greater in the district courts than in the dangerous deserts that we have had to explore and measure."[47] In Durango, Rafael García Martínez' report to Fomento that the judge was placing "all kinds of difficulties" in the way of his survey, "to the extreme of allowing that the existence of the company may be denied" provoked a strong denial from the judge. The company made a similar appeal two years later.[48]

Only in Tabasco and Chiapas, however, did surveyors complain consistently about lack of cooperation by local authorities. Bulnes Hermanos' confrontation with local officials in Tabasco is treated in detail in chapter 4. In Chiapas the Mexican Land Colonization Company resented the "constantes intrigues" carried out by the state government against the company for years, directed (in the company's opinion) by people who wanted the land it was surveying.[49] The company's charge of lack of cooperation by authorities there, especially in requiring individuals to present titles, has already been noted. Fomento routinely sent polite letters to officials in Chiapas asking them to be helpful but obviously considered the matter too delicate to risk any stronger action.

Similarly, when Policarpo Valenzuela complained in June 1888 that the federal judge in Tabasco, Simón Parra, was going out of his way to obstruct the survey, particularly by admitting new public land claims, Fomento discreetly asked for an explanation. The judge's reply—that he was only admitting claims for areas where Valenzuela had not yet begun to survey—was simply copied and sent to the surveyor. Several months earlier, the judge had delivered a sharp warning to the company to increase the pace of its surveys, because it had only begun work in the partido of Jonuta despite having designated many other zones. Valenzuela's defense was the lack of availability of surveyors, which the judge rejected on the grounds that Valenzuela simply was not offering to pay them enough.[50]

In January Valenzuela's representative at Fomento again protested that Judge Parra "palpably harasses [the survey] at every turn," partly by insisting on supervising the survey even in distant areas, where he should name a local magistrate to take control.[51] The judge was still accepting public land claims in April, Valenzuela complained again, complicating the operation and delaying its conclusion. Parra also kept appointing surveyors employed by the company to survey the claimed land, holding up Valenzuela's survey. Fomento again asked for a report from the judge, who took the opportunity to blister Valenzuela for dividing the whole state into six zones so big that it is "materially impossible for the survey to begin and continue simultaneously in every point or partido of the State." The company probably just wanted to monopolize the survey-contract business in Tabasco, he added. The judge's solution, he told

Fomento, was to require division into smaller zones, in which public land claims would not be halted until Valenzuela began work in them. Stopping all public land claims in the state just because Valenzuela designated the whole state for survey would be "un gran absurdo." Parra also accused the company of paying off public land claimants to get them to drop their claims, a practice of dubious legality.[52] The conflict was still brewing three years later, with the same judge still admitting claims in the "very center" of the survey, the company complained, and continuing to display "a marked animosity to the Company."[53]

THE GOVERNMENT'S ROLE IN INTERCOMPANY RIVALRIES

In practice the government appeared to have no consistent policy about whether to allow survey companies to operate in mutually exclusive zones. Exclusivity generally seemed to be the rule in Chihuahua, Sinaloa, Chiapas, and Tabasco, although companies in the first two states occasionally clashed over territorial rights. Rivalry seemed to be fiercest in Durango and Sonora.

Fomento's policy on the issue could change suddenly. In 1887 Asúnsolo's company complained that Ignacio Sandoval had just been authorized to survey in the Chihuahua cantónes of Mina and Victoria, two places that Asúnsolo had been contracted to survey. This led the judge to suspend Asúnsolo's operations, which had already begun in Victoria. In reply, the government reminded the federal judge that its authorization for Sandoval was *"sin perjuicio de tercero"* ("without prejudice to third parties") and that furthermore it never considered survey contracts to be "privilegios exclusivos," so there was no reason both could not continue their surveys at the same time. Yet within six months Sandoval's authorization had been limited to Mina, because Asúnsolo was already in Victoria.[54]

Unlike the Chihuahua contracts, which usually limited surveyors to certain cantones, those issued for Durango frequently overlapped, leading to numerous conflicts. In 1883, after Telésforo García pointed out to Fomento that his company had spent sixty thousand pesos surveying some 1.7 million hectares and did not want anyone else interfering, Fomento told the judge to keep other surveyors out of the partidos designated by García.[55] But the situation became so serious that four of the companies decided to merge their authorizations in 1884—Asúnsolo, Meza y Socios, Telésforo García, and the firm of Telésforo García and Jesús Meraz. The merger was initiated by Asúnsolo because of the "difficulties that the interests and pretensions of the rest of the survey companies were posing," which required action to "avoid the upheavals and obstacles" that resulted. Since all the concessions together covered nearly the whole state, the newly organized Compañía de Concesiones Unidas de Deslinde de Terrenos Baldíos en el Estado de Durango obtained the federal judge's permission to survey throughout the state. Subsequent surveys were conducted under the management of Asúnsolo's company.[56]

Fomento seemed to prefer letting the companies resolve for themselves the "upheavals and obstacles" created by the ministry's overlapping authorizations. When Joaquín Casasús opposed José María Calderón's survey of the pueblo of Buenaventura in 1890, Fomento first asked the latter whether he had arrived at *algún arreglo* (any arrangements) with Casasús; the answer was yes.[57] Asúnsolo's company clashed with García Martínez's when the latter was accused of invading land in Chihuahua already surveyed by Asúnsolo. A Fomento official found that both had surveyed "much of the same land" because of the haziness of the boundary between the two states. His solution: wait for the two firms to make a private arrangement. They soon did so, with Asúnsolo agreeing to withdraw its opposition to García's survey, in return for two thousand hectares in Durango.[58]

Sometimes the ministry's intervention was unavoidable, however. After Lauro Carrillo and Manuel Peniche's companies clashed in Sonora, Fomento mediated, and both sides agreed that Peniche would give Carrillo 40 percent of the one-third he was to receive in compensation for the survey of Sahuaripa district.[59] When Peniche's company was accused by Ignacio Gómez del Campo's of invading land already surveyed by the latter, Fomento investigated and found that Peniche had not only surveyed 57,142 hectares of land already measured by (and partly titled to) Gómez del Campo in Chihuahua but had also surveyed 60,853 hectares surveyed by and titled to Plutarco Ornelas. Peniche had to return the title he had gotten for land in compensation, which was reduced by the one-third of the land already surveyed by the other two companies.[60]

Similarly, when the Mexican Land Colonization Company's claims to land between the Yaqui and Mayo rivers were contested by Carlos Conant, who held an irrigation contract, Fomento intervened and mediated a settlement under which the companies would share the land. In this case, each one's contract had been so worded that the terms were irreconcilable without a new understanding.[61]

Fomento also mediated an agreement in 1887 between Luis Huller and José María Becerra after the latter angrily pointed out that Huller had designated for survey the entire state of Sinaloa, which if allowed to stand would nullify his own contract. The deal allowed Huller to survey baldíos everywhere but in Fuerte district, while Becerra would get Fuerte and the right to survey demasías there and in the rest of the state.[62]

Still, the hands-off policy dominated, the purpose apparently being to achieve the survey of the public lands in the shortest possible time, even if it led to intercompany conflicts, overlapping surveys, and titles whose claims overlapped. When a frustrated federal judge in Sonora tried to keep Huller and Olivares from surveying Ures district simultaneously (both had been authorized to operate there), Fomento overruled the judge, instructing him to let both companies work there simultaneously while taking care to keep them off the same land.[63]

Fomento even gave contractor Vicente Dardón permission in 1890 to resurvey land in Tabasco that he accused Policarpo Valenzuela's company of not having done accurately. The ministry warned Valenzuela that if Dardón's charges were true, "the responsibilities that may result will be demanded of you." Valenzuela's representative, calling Dardón's accusations "a solemn lie," pointed out that Dardón had an incentive to find errors so he could get one-third of the surveyed land. Fomento probably considered this a sound reason to authorize Dardón to rectify the survey. Since the work was "científico," the ministry replied, it scarcely mattered who did the resurvey. But after eighteen months, Dardón still had not submitted a single proof of his charges, a Fomento official noted, nor even indicated that his company had begun the resurvey, which ultimately was never conducted.[64]

Competition between the companies led to some sharp practices. Peniche accused Carrillo of collecting the titles from owners of various properties in Sahuaripa district just so Peniche would be unable to survey there.[65] Real-estate espionage was practiced by Andrés A. Quijano, who was authorized to arrange composiciones in Sonora's Magdalena district in 1890. His brother Fiacro went to work as a surveyor for the International Company, found out where the demasías were that the company intended to claim, then entered claims for them on behalf of Andrés. After International's lawyer, Emilio Velasco, reported the misconduct, the ministry invalidated Andrés's contract.[66]

CONCLUSIONS

In enforcing the laws and regulations governing the surveys, Fomento did not hesitate to move against surveyors whose behavior it considered detrimental to its goals of securing accurate surveys while maintaining peace in the countryside. When companies sought to enlist Fomento's support in disputes with local political authorities, the federal agency tended to avoid intervening no matter how much the company complained that its rights were being abused, which suggests that the central government was more responsive to local political conditions than to the sensitivities of its survey contractors. In conflicts between or among survey contractors, Fomento occasionally intervened to mediate but in general remained aloof from the struggle, preferring to let the contactors and local authorities resolve their disputes.

Even a state as devoted to economic development as that of Porfirio Díaz could scarcely be called the instrument of capital—or at least of that segment of the entrepreneurial elite represented by the survey companies. Chapter 4 illustrates that, aware of the threat to stability posed by attacks on the rights of nontitled occupants as well as legitimate landowners, Fomento and President Díaz frequently acted to protect those rights. This chapter has shown that the ministry in charge of the surveys generally enforced the contracts and required the companies to adhere to the legal and technical requirements it set. It disciplined surveyors for violations by nullifying titles, abrogating or suspending authorizations, and imposing fines.

The federal government's response to controversies that arose between the surveyors and local authorities reveals a more complex reality.

On the one hand, while the formalities of surveying relied heavily on the use of federal district judges and judges of inferior courts, ultimate control of the companies and their operations clearly was centralized in the Mexico City bureaucracy. The judges demonstrated little independence, and when conflicts developed the judges generally followed the cue given by Fomento. Evidence of intervention by state and other local officials is conspicuously absent in the survey company files, indicating that the federal government's right to dispose of the public lands was not even a matter of debate as early as the 1880s. The central government's authority was not questioned, and President Díaz was universally recognized as the ultimate arbiter. "The President of the Republic has decreed that. . ." was more than a legalistic phrase with which the federal executive responded to a persistent petitioner, whether survey company or landholder. On the other hand, when conflicts involving the companies and local officials such as federal judges did emerge, Fomento was often as not content to let the judges work out a satisfactory solution, even at the expense of the companies. Its reluctance to defend the surveyors in these cases once again reveals the central government's predominant commitment to other interests in the cause of economic development such as, in this instance, political stability.

Finally, we have seen that the government responded casually to jurisdictional conflicts among the companies—which in some cases appear to have been deliberately created by Fomento's practice of issuing overlapping permissions to survey. The government was also willing to tolerate the possible damage to survey company interests that resulted when landholders rushed to perfect pending public land claims. Both of these responses indicate that the federal government's top priorities remained the harmonization of property relations in the countryside. This goal would be served by the survey of the public domain in the shortest possible time and by encouraging public land squatters to submit claims, even at the expense of the survey companies' own interests. The companies were in most respects mere instruments, or as Porfirio Díaz called them, "cat's paws." The Porfirian state directed Mexico's transition to dependent capitalism with a firm hand, focusing, shaping, and policing the activities of entrepreneurs as it sought to balance conflicting interests and serve goals—in this case, reasonably accurate surveys and minimal class violence—that it judged beneficial to Mexico's long-term economic development.

4 Property Rights in a Modernizing Economy

Resistance to the Surveys and the Companies' Response

 This chapter assesses the impact of the surveys on the landholders in the six study states by analyzing the complaints raised against the surveyors by landholders, weighing the Díaz government's interaction with the protesters and the companies when conflict between them occurred, and presenting the companies' response. The evidence that supports the conclusions submitted herein is reported in Appendix B, and the evidence itself is exhaustive; that is, every indication of a protest by a landholder was recorded and noted by explicit reference to the event. Thus, all protests are summarized in Appendix B, with these exceptions: those arising in Sinaloa, where the volume of protests required a tabular form of presentation, and two cases that are reported in this chapter by way of illustration. Care was taken in all instances to report the response of the authorities to the complaints, as well as the companies' response.

A caveat unrelated to the quality of the sources is in order. The cases of usurpation reviewed here are limited to companies hired by the state to locate public lands. These particular survey companies were not the only entrepreneurs active in the disposition of the public land, or in surveying for that matter. As Table 1 indicates, more than seven thousand titles to public land were issued between 1878 and 1908 as the result of *denuncios* (claims) filed on public land, each of which required the submission of privately conducted surveys. Some 17 million hectares were distributed under the claims procedure, nearly as many as were given to the survey companies in compensation. The government itself sold about 52 million hectares of company-surveyed land classified as terreno nacional, more than the sum of all lands claimed and all lands given to the companies. Terreno nacional was often resurveyed when it was purchased. To the extent that any surveyor represented a threat to the land-tenure status quo, any kind of public land transaction might have generated protests against surveyors. Other usurpations occurred that had nothing to do with the disposition of the public lands. Growers, cattle ranchers, miners, real estate companies, and developers of all kinds undoubtedly infringed the property rights of many people, and the victims included other capitalist entrepreneurs as well as peasants. There were, in short, more opportunities for land usurpation during the Porfiriato than could ever be exhaustively studied by any historian working beyond the village or perhaps district level.

Before analyzing the allegations of land usurpation, we will first consider two related aspects of the problem. The first section will review some of the preventive measures taken by Fomento to obviate the threats to public order that the surveys could provoke by declaring—falsely or not—privately held land to be baldío. Then we will analyze the resistance to the surveys from the point of view of the companies themselves, who often regarded such resistance as nothing more than obstructionist and harassing behavior not only by landholders but also by local and federal government officials.

PREVENTIVE MEASURES BY THE STATE
TO KEEP THE PEACE

Some contracts contained stern language regarding the protection of land needed by villagers, even when the property lacked a title. The following provision was part of Andrés Quijano's composición contract for the Sonoran district of Magdalena in 1890: "The concessionaire will exercise extreme prudence in carrying out his operations, and in case of any difficulty with the pueblos," the contractor will report to Fomento so that the ministry could "make available every kind of facilities to them [pueblos] and give them every kind of guarantee, the said Secretariat dictating in every case the measures that it may judge necessary, and aiming to leave them in the calm, peaceful and legal possession of the necessary lots, in those cases in which they lack property titles."[1]

Similarly, shortly after the troops of the Primera Zona Militar, which included Sonora, won a victory over the Indians in 1886 under the command of Gen. Ángel Martínez, President Díaz asked for Martínez's views on the survey of the lands of the Yaqui River. The general, writing from Torín on the bank of the Río Yaqui, replied in May that any such effort in that direction would be premature, since although "it is certain" that the hostilities with the Indians have ended, "nevertheless peace has not been consolidated, to the point where these Indians are accepting their situation; and they may try to rise up again upon seeing that someone wants to confine them to a certain piece of land, since by custom they have lived many years independently, selecting at will the land that has pleased them the most." The land could be surveyed, he advised, as soon as the governors of the pueblos—whose influence is "decisiva"—allied with the government.[2]

As late as 1889 and 1891 the International Company, in its survey of Sonora, was told to stay away from land adjudicated to Yaqui Indians. Upon receiving permission to determine the location of twenty-five sitios given to the Yaquis, the firm was warned not to survey any land "that may be in the present possession of the pueblos of indigenous peoples, for which reason the Company will use much prudence in carrying out its operations."[3]

General Pacheco, the Fomento chief, circulated a notice to survey companies in 1883 stating that some of them were operating "in an unsuitable and even arbitary manner," infringing the rights of owners and individuals occupying baldío. The companies were reminded not to unduly upset such people, partly by making certain that the land about to be surveyed was designated in advance and was truly baldío, and that the federal judge's authorization had been requested. The objective was to eliminate any "reasonable motive for complaint."[4] Less than two years later, upset by the publication of a letter in two capital newspapers regarding the "alarm that reigns" in Chihuahua owing to the "abuses that some land survey companies are commiting" in declaring private property baldío (including one of Luis Terrazas's haciendas), President Díaz warned the companies that indemnification for damages resulting from such actions would be their exclusive responsibility if they did not adhere strictly to the law. A clarification was soon dispatched to federal prosecutors stating that the government was not asking the companies to respect any but "true property titles," because the government was just as eager to recover "the lands that have been usurped from it" as it was to protect private property. The companies, it pointed out, were agents of the government.[5]

Perhaps the most effective step taken by the government to protect landholders in the later years of the Porfiriato was to give occupants of baldío preference over survey companies (or any other claimants) in the disposition of the two-thirds of the surveyed land. This was especially important in Sinaloa, where Martínez de Castro's contract (later the Sinaloa Land Company's) specified the contractor's right to buy the two-thirds at a set price, provided that anyone occupying the land did not want to acquire it instead. As a result, Fomento usually sent a separate notice to the occupants informing them of their preferential right to buy any of the two-thirds whenever the survey company offered to buy it from the government. Most such offers were either turned down or ignored, but the ministry frequently took the additional precaution of asking the governor whether the land was occupied by persons who wanted to exercise their right of first refusal. In all of the post-1900 Sinaloa surveys, landowners bordering the surveyed territory were asked not only to give their consent to the boundary lines but also to the location the company proposed for its one-third of the land.

THE COMPANIES REACT TO RESISTANCE

The survey companies carried the federal government's warrant to measure and establish the boundaries of public land, and in doing so were free to choose the sites that produced maximum returns for their owners, subject to the approval of the local federal court. The companies were usually owned or managed by individuals associated in one way or another with the local elite, giving them extra clout in any conflict with landholders, particularly the poor.

Yet the surveyors often saw themselves as the victims of recalcitrant landholders, and they complained bitterly about obstructionist tactics sometimes directed against them by private landholders, local judges, and politicians. The companies' interactions with local government officials have already been covered in chapter 3. This section summarizes the most frequent grievances that the companies directed toward landholders. The archival records reveal a pattern of complaints to Fomento and federal judges about residents who refused to produce titles to land that they claimed to own; the submission of titles that were vague, ambiguous, or patently spurious; the initiation or resumption of public land claims under the 1863 law; and Fomento-arranged composiciones that excluded any right to compensation for the surveyors.

Residents who refused to present titles to land they claimed were a constant source of irritation, because they delayed the survey and also kept open the possibility that following the survey a court challenge would be successful if the plaintiff decided to produce his documents at that time. In Rafael García Martínez' survey of the partido of Durango in the state of Durango, such refusals constituted an "insuperable obstacle." Meza Hermanos y Compañía asked for, and received, two additional years to survey Durango because of the "infinite number of problems of all kinds that arise from the resistance offered by many of the proprietors whose land borders [the surveyed land] and which they occupy and hold on to illegally" without titles.[6]

Still, the hazards of resistance to a survey were sometimes threatening enough to persuade a company to let residents keep that to which they were not actually entitled. When the Mexican Land Colonization Company found landowners moving their boundary markers up to limits of already-surveyed public land, the company overlooked the larceny because "it was inconvenient to provoke these matters, which could have caused some risks. The Company believed it prudent to affect ignorance of them [and] accepted the proprietors' indications," since its surveyors had to work in deserted regions without sufficient protection.[7]

The collision of the two main public land laws of the time—Juárez's 1863 edict establishing a claims procedure and the 1883 survey law—was a constant theme in the archival record. The companies complained repeatedly that as soon as they designated the land to be surveyed before the judge, claims were filed against it by others; another common complaint was that long-neglected claims were suddenly reactivated after the designation. Twice in 1884 Fomento told all federal judges not to accept claims against land already designated for survey by the companies, nor claims on land already surveyed, because the latter was no longer subject to claim under the law of July 22, 1863, but instead under the 1883 colonization law. The warning had to be repeated to the federal judge for Chihuahua the following year, and again nationwide in 1887.[8] To protect the government's interests, however, Fomento (after consulting President Díaz) insisted that abandoned public land claims, if already surveyed by the claimant, were not to be included as baldío by a survey company unless it was impossible to locate them.[9]

The prohibition of claims against land already designated by a surveyor was an advantage that could be abused by the companies. Designated zones that were excessively large blocked land claims and surveys by other companies, retarding progress in disposing of the public land. By 1891 Fomento was trying to limit such zones to the size of a municipio.[10] Yet the Mexican Land Colonization Company's Emilio Velasco argued that the company should not have to respect pending land claims at all, even those filed before the company was authorized to survey. The claimants themselves should have to enforce their claims. The firm's only obligation is to publicly announce its plans so that anyone with a pending claim may present it. When the surveys began in Chiapas no one paid any attention, Velasco said, because the practice in that state had always been to claim land, take it, and then indefinitely drag out the claims process to avoid paying for the land. Property holders in Chiapas assumed they could continue this subterfuge even after the company arrived on the scene, he said. Such claims were done in bad faith and the company need not respect them, Velasco added, pointing out that some claimants were not above changing boundary markers to take in better quality land.[11]

Individuals who claimed land that to a surveyor was indisputably baldío undermined the entire economic rationale for the company's existence, because unless a company discovered baldíos, it would not be able to secure the only compensation to which it was entitled. The companies frequently pointed out to Fomento that many years—perhaps decades—had passed, during which a landholder made no effort to perfect his title, despite the existence of legislation that often made it easy for him to do so. Then the surveyors appeared, and the companies had to fight spurious claims. As noted in chapter 2, this is precisely what the government intended when it adopted the survey company policy. By threatening to turn over imperfectly titled land to the government or a survey company, the state provided landholders with an incentive to perfect their claims. Some of the companies did not appreciate this aspect of Porfirian land policy.

In Chiapas, for example, it was not until after the Mexican Land Colonization Company finished its surveys that various property holders decided to resume action on public land claims initiated long before the survey—one of which dated from 1840. In most cases they had done nothing but file the claim, and had never bothered to survey the land or otherwise follow through on the action. The company informed Fomento that because its survey sufficiently increased the value of the land, property holders had been prompted to resume their claims. It took two more complaints before Fomento finally told the local judge to suspend all action on public land claims in areas already surveyed by the company.[12]

Jesús Valenzuela's company was incensed by Fomento's admission of composiciones in Cantón Iturbide without verifying the individuals' real right to the land, as noted earlier. Antonio de Cárdenas, representing the Compañía Deslindadora de Terrenos Baldíos of San Luis Potosi, complained to President Díaz that when property owners arranged a composición directly with Fomento, any survey company work was suspended. Such decisions made it impossible "to define the property or to know what belongs to each person." As a result, "in one way or another the will of the property owner endures, and we can never arrive at a clear solution." Citing the "tenacious opposition that a certain number of united property owners can produce (and whose association forced previous companies to go out of business)" he asked President Díaz:

> Sir, why do some property owners resist the measurement of their land and the presentation of their titles? The survey and establishment of boundaries costs them nothing. It is a convenience to them to vindicate themselves of any possibility that they have property that belongs to the Nation or to other individuals; and nevertheless

they reject it. There are property holders who request the survey of their properties, and they do not refuse to present their titles because they have nothing to hide.[13]

Joaquín Casasús, who held a survey concession for Durango, objected to the uncompensated costs imposed on the companies by the recalcitrance cited by de Cárdenas:

> In fact, without the survey companies, the property owners would never appear before the Government to regularize their property; because in spite of the advantages that the laws have offered them and the preferences that have been created in their favor, they have not taken care to avail themselves of the benefits of those laws and of the protection of those preferences. Today when the threat exists that a Company is going to measure their land, when the frauds that have been committed are discovered, they rush to appear before the Government and claim that they ought to enjoy all the benefits that have been offered to them.

The government's aim, Casasús continued, was therefore fulfilled— "the proprietor regularizes his property, and is no longer a simple squatter." But the survey company, meanwhile, has spent money on the survey, which is "without a doubt unjust." As soon as he surveyed a finca, Casasús complained, Fomento arranged a composición with the resident, "oblivious to the fact that without the effort that I have put forth and the expenses that I have undertaken, never would he [the property owner] have agreed to secure his property, guaranteeing it against any attack."[14]

The government replied that his contract provided for just such compensation, and that he would get part of the revenue received by the government for the composición. A lawyer like Casasús scarcely needed to be reminded what his contract said; it is likely that the provision simply did not allow for "adequate" compensation.[15]

RESISTANCE TO THE SURVEYS AND LAND USURPATION

The quantity and variety of complaints that turned up in Fomento's survey company files, as reported in Appendix B and in this chapter, indicate that rural property holders, whether latifundistas, rancheros, or peasant communities, typically had the freedom to register their opposition, and also found it within their means to do so. The 133 protests uncovered in the archives indicate that formal resistance to a survey was not an extraordinary event.[16] Neither local authorities nor the survey companies were holding guns to the heads of rural landholders to prevent them from complaining. Of the complaints reviewed in this chapter, none mentioned a threat of violence to enforce compliance with a survey by one of the concessionaires.[17] Precisely the opposite tendency was seen—the companies and local authorities complained that property holders, including peasant and Indian communities, threatened to revolt or to drive the surveyors away, and sometimes so intimidated the surveyors that they fled for their lives.

As mentioned earlier, the cases reviewed in this chapter were exceptions. The vastly more numerous cases in which property holders presented evidence of ownership that was accepted by the companies—who thus (according to the records) respected their properties—are not included herein.

The judicial records indicated, for example, that the companies routinely respected land set aside as ejido or fundo legal, and that conflicts over such properties were exceptional, notwithstanding the constitution's prohibition of corporate ownership of land, including ejido. When Jesús Valenzuela's company surveyed Cantón Guerrero in Chihuahua in 1884, the surveyor noted that although the pueblos of Ariziochi, Tomóchi, and Pichachic "have no documentary titles, it is undeniable that each one of them has the right to a *sitio de ganado mayor*" (each of which was equivalent to 1,756 hectares).[18]

In José Valenzuela's survey of the directoria of Choix in Sinaloa, the engineer noted that he was leaving the municipality some land, despite the constitutional prohibition,

> considering that the spirit of the Federal Government upon undertaking the survey of the national lands has not been to disturb the farmer who with slight legal foundation is established on some site; but, on the contrary, has instead had to protect the acquisition of the property and to disseminate the yield of uncultivated lands, and desiring, on the other hand, to remove the difficulties that arise if the present holders of such lands were to be dispossessed of them.

He also noted that the federal government had appealed for flexibility in such cases, confirming ejido grants and even giving land as ejido to pueblos that never had any, under the condition that the inhabitants immediately divide it among themselves, informing Fomento so the latter could issue the necessary titles. Even though the inhabitants of Choix "have not a single document that would protect a scrap of land, nor would sanction the remission of land to its residents," he left them 576 hectares, enough "space so that the people could breed and run their cattle without doing any damage."[19]

Piratic behavior by the survey companies, though not unheard of, was usually unnecessary for two reasons: plenty of genuinely vacant public land was still available, and the companies were free to demand proof of clear legal title from any property holder whose land they desired—a powerful weapon in a country where customary use and dubious documentation were so common. What is surprising, perhaps, is the frequency with which rural inhabitants were able to produce such evidence. In any case, as indicated above, survey concession holders complained often enough of abuse in the countryside to suggest that they lacked the unconditional authority frequently attributed to them in the historical literature. Besides, as profit-making concerns they had a natural incentive to avoid a fight in order to hold down costs, even if that meant letting property holders steal their land, as the Mexican Land Colonization Company discovered in its survey of Chiapas, as noted previously.

More problematic is the question of the outcome of those complaints that did arise. Only occasionally could a clear victory for either side be detected. In most cases the result was inconclusive. Sometimes the protests themselves were so vague that they scarcely admitted a conclusion, as in the case of the eleven Tarahumara Indians who simply asked the judge to protect their land. Like many other protesters, they made no specific accusation against the company but simply asserted their right to continue enjoying the property they held, which the surveyors may or may not have been attempting to designate as baldío.

Nevertheless, I have attempted to analyze the protests, the results of which are presented in Tables 13, 14, and 15. Except for the surveys by Luis Martínez de Castro and his successor, the Sinaloa Land Company, the parties lodging the complaint were identified as either *entrepreneurs* (Table 14) or *communities* (Table 15). These terms are shorthand for two distinctive types of rural producers. Entrepreneurs include hacendados, miners, ranchers, and other individuals whose activities, as described in the archival sources, appeared to qualify them as capitalists. Communities were protesters whose tie to the land appeared to be shared collectively and was rooted in tradition. The latter group most commonly included those petitions signed by numerous *vecinos* (residents) of a pueblo, or by groups of *indígenas* (indigenous peoples). This classification is an admittedly crude attempt to make use of what little information the records supplied about the protesters, whose class interests and access to resources must have varied greatly. This problem is somewhat analogous to the quality-of-land problem raised above, for people, like the land from which they make their living, come in all varieties. The scarcity of

Table 13 Opposition to Surveys by Two Major Contractors in Sinaloa

	Sinaloa Land Co.	Martínez de Castro
Total lots surveyed	101	48
Surveys rejected	28	11
Proportion of rejections to total lots surveyed	28%	23%
Surveys opposed	35	14
Proportion of surveys opposed to total	34%	29%
Surveys rejected that were also opposed	20	5
Proportion of opposed surveys rejected	57%	35%
Of all opposed in court, total withdrawn or abandoned	5	4
Proportion of court oppositions withdrawn	14%	28%

Source: Archivo de Terrenos Nacionales, Sec. de Reforma Agraria.

qualitative data, while far more serious in the case of land (because it is almost completely nonexistent) than in the case of the people associated with the land, is not crippling, however. In almost all cases enough information usually appeared in the correspondence and other documents to make it possible to judge the status of the protester in terms of his participation in the emerging market economy.

Because of the great volume of surveys conducted in Sinaloa by Luis Martínez de Castro and his successor after 1900, the response to these surveys is analyzed differently from those of earlier surveyors in Sinaloa, and from those of surveyors in other states. The difference is twofold. First, no effort was made to distinguish entrepreneurs from communities, both because the distinction was more elusive late in the Porfiriato, when these two firms were active, and because the great number of surveys by these two concessionaires in Sinaloa (149) made it less practical to collect such data. Second, instead of classifying Fomento's response to a particular complaint as either for the company or the protester, the outcome of Fomento's ruling on the survey itself (i.e., whether it was approved or rejected) was recorded. Since these two companies' surveys took place at least ten years after most of the others were done, separating

Table 14 Opposition to Surveys in Six States by Entrepreneurs

State	Total Protests	Resolution					
		For Protester	For Company	Nego-tiated	Dropped	Referred	Unknown
Chihuahua	32	13	1	15	0	0	3
Sinaloa[a]	7	3	0	0	0	0	4
Durango	4	0	1	3	0	0	0
Sonora	4	0	2	0	1	1	0
Chiapas	3	0	0	0	0	0	3
Tabasco	3	2	0	1	0	0	0
Totals	53	18	4	19	1	1	10

Source: Archivo de Terrenos Nacionales, Sec. de Reforma Agraria.

Note: This table and Table 15 exclude three cases—one in Chihuahua and two in Sinaloa—in which the protesters could not be identified as either "entrepreneur" or "community." See text for definitions.

[a]Excludes Sinaloa Land Company, Martínez de Castro, and Sandoval-Pacheco. Data on the first two are in Table 13; the last named company's surveys were nearly all composiciones and are excluded for that reason.

them does less violence to the overall analysis than would otherwise be the case.

With these distinctions between post–1900 Sinaloa and the rest of the study area in mind, let us review the data presented in the three tables. Eighty-three protests of some kind are reported in Tables 14 and 15. Approximately two out of three (64 percent) were submitted by entrepreneurs, and nearly all the others by communities. (Two protesters could not be adequately identified.) Among entrepreneurs, about one out of three protests (34 percent) were resolved in favor of the complainant. Fomento favored the company in four others (8 percent), direct negotiations solved the conflict in nineteen (36 percent), and the resolution of about one in every five (19 percent) could not be determined. The rest (4 percent) were dropped or referred to other authorities (100 due to rounding). Of those submitted by communities, about two out of five (43 percent) were resolved by Fomento on behalf of the protester. Only one (4 percent) was negotiated independently by the company, three (11 percent) were referred by Fomento to another authority (usually the court), and the resolution of the remainder (43 percent) could not be determined.

Of the forty-nine surveys opposed in Sinaloa after 1900, Fomento rejected about half (51 percent), at least partly (and usually solely) because of some defect identified by the protesters.

Table 15 Opposition to Surveys in Six States by Communities

State	Total Protests	For Protester	For Company	Nego-tiated	Dropped	Referred	Unknown
			Resolution				
Chihuahua	9	3	0	0	0	3	3
Sinaloa[a]	5	2	0	0	0	0	3
Durango	3	2	0	0	0	0	1
Sonora	0	—	—	—	—	—	—
Chiapas	9	4	0	0	0	0	5
Tabasco	2	1[b]	0	1	0	0	0
Totals	28	12	0	1	0	3	12

Source: Archivo de Terrenos Nacionales.

Note: This table and Table 14 excludes three cases—one in Chihuahua and two in Sinaloa—in which the protesters could not be identified as either "entrepreneur" or "community." See text for definitions.

[a]Excludes Sinaloa Land Company, Martínez de Castro, and Sandoval-Pacheco. Data on the first two are in Table 13; the last named company's surveys were nearly all composiciones and are excluded for that reason.

[b]In this case the surveyors were driven away violently by the protesters.

The quantitative analysis discloses Fomento's reluctance to side with the survey company in a conflict with a property holder. When the ministry did so before 1900 it was invariably in a case involving an entrepreneur. In not one case involving a community did Fomento take the side of the company in a conflict. On the other hand, Fomento appeared less likely to flatly resolve a conflict on behalf of a community protester than an entrepreneur. In cases involving communities, the proportion of cases whose resolution was indeterminate was more than double that for entrepreneurs. In other words, the record is more often silent about the result of cases involving communities than for entrepreneurs, though in Sinaloa after 1900 the government seemed more inclined to rule on behalf of the protesters.

The cases in Appendix B show that the federal government frequently sided with the protesters against the companies, sometimes suspending surveys and revoking titles, at other times forcing the companies to respect occupied land even when it was not properly titled. Local authorities such as judges and prosecutors also acted on behalf of property holders whose land was threatened by the companies. Well-to-do land-owners were obviously the most likely to receive such treatment, but poor farmers and peasant communities were also accorded protection.

Sometimes the authorities in Mexico City or in a local jurisdiction were able to protect landholders indirectly for years simply by not acting on a survey or by refusing to issue a title to the company pending the outcome of a protest.

Disputes between landholders and surveyors also seemed to be over truly contentious issues. When clashes occurred, the complainants were often the first to acknowledge the absence of clear title to the land they claimed. Even in the record of Sandoval's composiciones in Sinaloa, which bore all the signs of an unscrupulous operation, his victims were undoubtedly occupying vacant public land.

To illustrate these trends, two cases in which landholders charged survey companies with usurpation of property will now be discussed. These cases were selected in part because the targets of the charges in both instances were precisely the sort of modernizing, industrial-oriented enterprises that the Mexican state was seeking to attract and nurture. Yet as we shall see, popular resistance to the surveys—in the first case from among a mixture of both entrepreneurial and community sources, and in the second case from largely community sources—triumphed over the companies in the form of successful appeals to the state for relief.

The Case of Bulnes Hermanos in Tabasco

No survey company appeared to generate such fierce, violent, and extensive opposition among local authorities and property holders as did Bulnes Hermanos in Tabasco. The San Juan Bautista merchant family was scorned for its Spanish nationality as well as its wealth and the privileges it enjoyed as a holder of Fomento concessions. In return for the resentment it created, and for the extra costs of resurveying and litigation that were incurred as a result of the opposition, the company had little to show for its survey contracts; the record contains only one title (for 12,281 hectares) issued to the company in Tabasco in compensation for a survey—and that one was nullified by a local judge in 1895.

Much of the popular opposition to the company resulted when a federal contract that was not even for surveying was confused with the survey contract, which made the company's job much harder. Its difficulties were compounded by a record of either technical ineptitude or deliberate dishonesty that few other contractors could rival. The firm's troubles started when it began work under a federal contract to build a pier on the Río Grijalva and offices in San Juan Bautista in connection with its obligation to establish colonists in Tabasco under a separate survey contract it acquired from Federico Méndez Rivas in 1884.

Bulnes was accused by the editor of the *Boletín Municipal* of seizing part of the city's fundo legal for the construction project, and of taking land the city had already sold to others. Four inflammatory handbills were circulated in December 1884, alleging that Bulnes Hermanos, "powerful merchants," were directly attacking the city itself by trying to obtain riverfront land for a private pier that would monopolize trade and bring contraband into the city. Meetings attracted six hundred people to protest President Díaz's issue of a title to the land.[20]

Matters came to a head when the city's *ayuntamiento* (city council) met in an extraordinary session on December 21 to hear a petition by five hundred individuals who presented themselves "en mucho órden" (to quote the official account of the ayuntamiento). They called attention to the "unheard-of plundering of a great belt of land" by Bulnes on the left bank of the Grijalva that was part of the city's fundo legal and therefore inalienable. The city did not need a pier, especially for colonization, since that would not only take many years but would only benefit Bulnes. The petitioners asked the ayuntamiento to seek the nullification of Bulnes's title, which the body voted unanimously to do. It sent a representative to the president, since "we would never doubt the upright approach of the worthy General Porfirio Díaz," according to one of the organizers of the protest.[21]

In January 1885, Tabasco Gov. Eusebio Castillo appealed to President Díaz on behalf of the ayuntamiento against the company. The latter, fearing that the governor's envoy would "distort the facts," sent its own representative, Federico Méndez Rivas.[22]

The company blamed the events of December on "people of perverse intentions" who riled the populace to such an extent that bloodshed was avoided only by the timely dispatch of a picket of federal soldiers. Bulnes decided that a postponement of its request for judicial possession of the riverfront land was in order. The company complained to Fomento that the hue and cry raised by opponents in San Juan Bautista was echoed in its survey of baldíos in the countryside, but for different reasons. With some justice, the firm acknowledged, the survey was greeted with alarm because the baldíos were being occupied under provisional titles issued by the state of Tabasco before their disposition became an exclusively federal matter (see note 72 of Appendix B). The present-day occupants based their right of ownership on those documents. This and the belief that the government had given the company "a twenty-meter belt along interior coastal ways" made it impossible for the surveyors to work, the company said.[23]

Bulness Hermanos appealed for Fomento's "moral and material support" and for permission to hire foreign surveyors to replace the local surveyors who were frightened off by threats to their lives. When the jefe político of the municipalidad of Tenosique saw the surveyors coming, "he armed people from the town and issued death threats" against the surveyors, who fled in terror. At Fomento's request the governor of Tabasco told the jefe político of Balancán department to try to avoid the use of violence by landholders of Balancán against the surveyors.[24]

Meanwhile, President Díaz sympathized privately with the beleaguered merchant family, telling them in December 1885 that "I am pained by the difficult situation in which you find yourselves." The president added that he hoped Bulnes achieved the remedy he was seeking by means of his complaint to the *comandante militar*, and he promised to write to the governor, as Bulnes requested.[25]

The local authorities, however, did not become noticeably more tactful in their treatment of the company. In April 1886, Canuto Bulnes complained to Díaz of $160 in fines he had to pay in Tenosique; it wasn't the money that bothered him so much as the way it was collected—the authorities imprisoned both him and his son Antonio until the fines were paid. The fines had been levied against the business's steamship *Asturias* for violating a municipal ordinance. Bulnes accused Emeterio Fabre, jefe político of the partido of Balancán, of being behind the campaign against the company.[26] The president was no supporter of Bulnes, if indeed he had ever been, and in 1888 he confided to the governor that regarding the Bulnes contract, "I recommend that you be careful with this case, since it appears to be injurious to the Country."[27]

Bulnes did not get to keep the title to 12,281 hectares it received in Balancán in compensation for its survey of the land between the Usumacinta and Chacamax rivers. In 1894, seven years after the survey, Ismael España of Tenosique told Fomento that the area titled to the company was never surveyed by it, "in spite of the fictitious operations that they appear to have carried out." Fomento forwarded the allegation to the federal district judge in Tabasco. A few months later Fomento heard that the company was invading land held by the pueblo of Usumacinta. The local federal prosecutor was asked to take action to recover the land titled to the company in Balancán, because their surveys were defective and the firm was illegally cutting timber and sowing crops. He was told to determine whether there was any fraud.[28]

The crowning blow to the company's survey was administered by a surveyor employed by the Comisión de Límites, which was charged with fixing the border between Guatemala and Mexico. His conclusion: "if by chance Messrs. Bulnes carried out any field operations to do their survey, they must have been done in a very limited area." Fomento instructed federal authorities in Tabasco to recover the title issued to the company and to demand an indemnity for damages, including $5,631 owed the

Comisión de Límites for its checkup of the Bulnes survey. President Díaz finally decreed that Bulnes would have to resurvey the land under the condition, among others, that the company respect land given "as ejidos to the pueblo of Usumacinta." Bulnes also had to renounce all rights to land he surveyed in Chiapas, except for 225 hectares at the mouth of the Chiniquija creek. Under the final decree, the company would not have to pay the $5,631.[29]

When Bulnes resurveyed the area between the Usumacinta and Chacamax rivers in 1900, Manuel S. Rodríguez opposed the survey, claiming his land had not been respected. The federal district court for Tabasco rejected his claim, but it was upheld on appeal in 1905. There was no need for Rodríguez to have rushed to court, however—the resurvey was not approved until 1913, and the required physiographic report was not submitted by Bulnes until 1920. None of the resurveyed land, totaling some 81,000 hectares, had yet been alienated as late as 1920.[30]

The Case of José María Becerra in Sinaloa

Probably the longest-running land dispute between a survey company and a community was sparked by Becerra in 1889 when his company found what it claimed were 3,432 hectares of demasías not accounted for under two titles—a colonial-era document issued to "los naturales" of the pueblo of Navolato (population in 1907: 1,884) in Culiacán district, and a second one issued by the prefect of Culiacán in 1868. The company allotted the pueblo 51 hectares for its fundo legal, then asked for 1,144 as its one-third and for the rest in purchase. Fomento declined the requests initially, saying it had to study the legality of the prefect's title. Then a lawyer representing three individuals "and other owners" of land in Navolato opposed the transfer of the demasías to Becerra on the grounds that his clients held the land under the title issued by the prefect, who acted in accord with the Ley Lerdo. (This 1856 statute prohibited civil institutions such as municipalities from holding real property, and required its sale.)[31]

A *juicio de oposición* (opposition proceeding) was initiated by the plaintiffs in August 1889. Meanwhile, a Fomento official's finding that the prefect acted legally was apparently ignored by the decision-makers at the ministry, who issued a title two years later to Jesús Almada (at Becerra's request) for all 3,432 hectares of demasías.[32] By 1898 Jesús and Jorge Almada had built a US$2 million sugar refinery and distillery that employed more than nine hundred workers in Navolato, an area that was still "the center of a wilderness" when Almada acquired his title.[33]

Six months after Almada got his title, Fomento received a letter from a vecino of Navolato that resulted in a decision by President Díaz to suspend Almada's right to take possession of the land. The vecino, Pomposo Medina, who also identified himself as the representative of "the other Indians" of Navolato, said the residents had originally protested the survey because the pueblo no longer held land in common, each family having the right to its own land under the titles issued by the prefect. They dropped their suit against the survey only after the company promised it would respect their land. As a result, Almada's title was issued. But the vecinos were outraged when the surveyor, acting in connivance with the mayor (who also happened to be the Almadas' *mayordomo* [majordomo]), located the demasías "precisely in the place most affecting our small interests," taking "a great number" of our "enclosures, houses and small farms." President Díaz responded by suspending Almada's judicial possession of the land until judgment could be rendered by the "competent authority." The president did not know the placement of the land when he had signed the title, nor the "irregularities that appear today in the map," Fomento told Medina.[34]

The Almadas' lawyer replied to Fomento that not only did he represent the sugar refiners but, to prove the groundlessness of Medina's complaint, fourteen "Indians of Navolato" as well. He argued that when the prefect divided the communal land in 1868, he gave away three times more land than the pueblo actually owned. The error was compounded ten years later when a new prefect divided the land yet again and handed out even more than in 1868. Furthermore, the land was divided not among the vecinos of the pueblo but among thirty-six "favorites" who "arbitrarily picked themselves." According to the firm's lawyer, only a tiny fraction of the distributed lands was actually marked off and cultivated as required by law.[35]

Fomento took the dispute seriously enough to send its own surveyor to Navolato. In a forty-nine-page report, Ing. Enrique Hijar y Haro concluded that the distribution of land to vecinos in compliance with the Ley Lerdo had been "absurdly impracticable," because the distribution was never conducted and most of the land at the time of his report had been neither cultivated nor marked off. After Almada acquired title to the land, the vecinos hired a surveyor who made a map that was a complete fake, serving only to "cheat the Government and create problems." Hijar y Haro's report included a detailed chart of that survey's numerous errors, and recommended yet another survey and a new distribution of land among the vecinos.[36]

By 1897, the construction of his refinery and distillery notwithstanding, Almada was still unable to take full possession of the land under dispute because of the vecinos' opposition. He proposed letting Hijar y

Haro undertake a new survey that would establish the boundaries of each plot once and for all and determine the quantity of demasías so that a settlement could be made with the vecinos. At the same time, Fomento discovered that Luis Huller had surveyed Navolato in 1887 and found 50,578 hectares of baldío and demasías. The survey had been rejected, in part because Huller had not left a fundo legal for the town, and because of the titles issued by the prefect. The ministry had then approved Becerra's survey, despite the fact that Huller's was still pending, thus making Becerra's inadmissible, a Fomento memo concluded. Then another complication appeared. Manuel Martínez del Río, who had acquired rights to the survey concessions of both Huller (for Culiacán) and Becerra (for Navolato), demanded compensation for both surveys.[37]

By 1907 the Almadas (now doing business as the Almada Sugar Refineries Company, S.A.) had still been unable to obtain full possession of the 3,432 hectares of demasías to which they had received title in 1891. But by this time they had also satisfied themselves that the vecinos were right all along—the demasías "were placed on land and lots that were perfectly titled," the Almadas' lawyer acknowledged. To "avoid problems," he proposed to Fomento that his company either be allowed to exchange the demasías for land that was "really free" elsewhere in Culiacán, or that the government return what the company paid for the land and reimburse its legal expenses. The government countered that it would reimburse the company for the one-third of the land that had been owed to the original survey contractor (Becerra), but that Almada would have to recover the value of the other two-thirds from Becerra, since the surveyor was to blame for the "bad designation and bad measurement" of the land. They compromised in January 1907. The title was nullified, Fomento agreed to return the $1,715 in bonds that Almada paid for the two-thirds, and Almada remained free to lodge a claim against Becerra for other damages. It was a complete victory for the vecinos of Navolato, who with the intervention of President Díaz had managed to prevent the company from taking possession of the land until the Almadas could be satisfied that the land was legally titled.[38]

CONCLUSIONS

What stands out above all in the quantitative analysis reported herein is the slight proportion of complaints to the total volume of land surveyed and to the number of individuals whose property bordered surveyed land. Unfortunately, it was not practical to quantify the number of property holders involved in each survey. Nor was it possible to add up the amount of land disputed by property holders, because the complaints rarely quantified the area in question. (To do so would have required a survey, which many property holders wanted to avoid.) In comparison to the overall scale of public land surveys, it seems clear that few landholders complained.

Sinaloa was the one state where the proportion of complaints was relatively high—one out of three surveys were opposed after 1900, as Table 13 indicates. While the other states were surveyed in the 1880s and early 1890s, most of the baldíos in Sinaloa were surveyed after 1900. Because the surveys were done late in the Porfirian period, well after the Northeast had begun its prodigious economic expansion, land values had risen sharply and the possession of land was more likely to be contested.[39] Moreover, Sinaloa's population density greatly exceeded that of the other northern states in this study, and had increased 25 percent in the last fifteen years of the Porfiriato.[40]

There is simply no evidence that the survey companies extensively usurped worked or titled land. On the contrary, the companies and the state usually respected the landholdings of both peasants and big landowners, and often set aside land for village ejidos. Property holders who did lose land because of the surveys usually did so because they were unable to prove that they owned the land in question, or that they had even possessed it, which would at least have qualified them for a substantial discount if they wanted to buy the land as baldío from the government. Some refused to cooperate with the surveyors at all.

Assuming that the companies behaved like vandals, expropriating land at every opportunity, historians not averse to *post hoc* reasoning extrapolated further and inferred peasant discontent transformed into revolutionary violence. Not only is the evidence of large-scale usurpation scarce, but most of the survey companies were compensated for their work some twenty years before the revolution erupted.[41]

Rural inhabitants of Mexico in the Díaz era actually benefited from the state's determination to base landownership strictly on legal documentation. Surveys of the public land, by establishing clear boundaries and making it possible to identify those limits unambiguously in a legally enforceable document, added value to land.[42] Even in the absence of a title, evidence of continuous possession was often accepted under the right of prescription. These benefits of the surveys can be measured by the increase in the price of land after it was surveyed.[43] Knowledge that property rights in land were exclusive and clearly defined attracted new investments and provided an incentive for all kinds of landowners to make improvements and intensify capital investments, an essential precondition to the spectacular economic growth of the Porfirian age.

5 The Impact of the Surveys on Land Concentration and Values

One of the most prominent themes of the historical literature on the survey companies could be summed up in the single word *monopolization:* The government's survey company policy further increased the concentration of landownership during the Díaz era, and thus contributed to the outbreak of the 1910 revolution. Since that literature is reviewed in some detail elsewhere in this volume,[1] here our task will be confined to testing the monopolization hypothesis. The claim is extremely difficult to test, since it requires precise data on the number of owners and the quantity of land they held before and after a survey in a given region. But in the absence of one of the conditions necessary for such a study, namely an accurate description of property boundaries both before and after the survey, the data needed for a rigorous test of the monopolization hypothesis are by definition unavailable, or nearly so. Nevertheless, a tendency or pattern for the companies to either dispose of their newly acquired land or to keep it can be determined within the limits of this investigation.

This analysis is therefore confined to the destination of the one-third of the surveyed land that the companies received in compensation for their labors. The two-thirds turned over to the government were available for sale as terreno nacional, of course, and no effort was made to determine how that land was distributed. However, given the size and geographical scope of the six-state sample of contractors' land that will be analyzed, it is reasonable to infer an analogous disposition for the government's two-thirds, especially because the government's share was contiguous to the companies' and became available for sale at the same time.

Evidence that the companies subdivided and transferred their shares of the surveyed land would contradict the claim, which has not yet been substantiated, that surveyed land was monopolized by the companies and that the survey policy of the Díaz government therefore increased land concentration. This chapter analyzes all the data found in the files of the Archivo de Terrenos Nacionales, as well as in notarial documents cited in those files, relating to the transfer of survey company land in the six states. The evidence provided here is exhaustive—that is, every case found in which survey company land was subsequently transferred is reported.[2] References to the transfer and subdivision of land were often found in letters to Fomento that requested copies of titles or clarifications of boundaries, and sometimes in complaints submitted by individuals that accused the new owner of usurping their land. In every case where a reference was made in Fomento's files to a notarized document dealing with the transfer of survey company land, that document was sought in the notarial archives of the Federal District. As will be seen, land sold or otherwise transferred by the companies was almost always distributed to more than one recipient.

In this chapter evidence will also be presented that the surveys added at least 50 percent to the value of land. The interpretation of the surveys' effects on both land concentration and land values that concludes this chapter will consider how the data relate to the rural discontent that sparked the revolution of 1910.

LAND CONCENTRATION

In Chihuahua, information about land transfers conducted by two of the largest companies operating in the state indicates that the property they acquired was rapidly subdivided and resold.

Ignacio Sandoval's concession for Chihuahua was owned by Sandoval, Concepción Sánchez, Andrés Salazar, Guadalupe Gutiérrez, Mariano Lachica, and Ponciano Falomir. They agreed that of the land to be given the company in compensation, two-thirds would go to Sandoval and one-fifth of the remainder to each of the other five. The partners also had the right to receive half "of the portion that corresponds to them of the lots adjoining or annexed to the rural property that they possess or where they may select." If any land was sold by the company, the product would be divided among the owners according to the above formula.

In 1889 Sandoval sold his partners 193,843 hectares in Mina included in his title of January 27, 1889, for $2,533; each of the five men would get one-fifth of the land (38,769 hectares). Sandoval's surveyor, William P. Morrison, was transferred Mina land in 1892; he sold his to Andrés G. Urquhart, who sold it to Seth B. Orndorff in 1906; George Ferguson and James Monk of Leesville, Louisiana, bought it from Orndorff in

1910, and later claimed they owned 15,000 acres (not hectares). In 1903 the Pine King Land & Lumber Company bought 209,111 hectares from Sandoval in Mina and Victoria. Sandoval's family sold more of their land to Long Hermanos and to José D. Knotts in 1907. Sánchez's widow sold Juan Chávez González the 19,384 hectares she still held, in 1920.

Mariano Lachica, one of Sandoval's partners, bought 7,024 hectares in the early 1890s from Morrison, the surveyor, and from another person. Lachica, defending his right to his land after the promulgation of the 1924 decree nullifying all titles issued under the concession, claimed that Sandoval's land, far from being the object of monopolization by a single individual, as charged in the decree, had given birth to "a true colonization, . . . since they have been divided among various co-proprietors, many of whom, . . . have no other assets."

In 1928 J. F. Carranza, a Fomento official, noted that among those who bought land from Sandoval's concession was José Félix Cano, who paid $675 for 2,632 hectares in 1889, land that was still in his family today. He added that they were "small farmers" and their case was similar to many others, not just in the Sandoval concession. Carranza believed that such smallholders should be protected from expropriation of former survey company land.[3]

When Felipe Arellano left the Valenzuela partnership, his payoff was 52,680 hectares of surveyed land in Cantón Camargo.[4] When Celso González left the company in 1888, he received as a return on his investment 334,594 hectares in Cantón Meoqui and another 195,985 in Ojinaga.[5]

In 1887 the company sold some of the 200,023 hectares it acquired in Cantón Rayón to the El Refugio Mining & Milling Company, and a few years later sold 12,190 hectares to a rancher for $3,475.[6] Valenzuela sold the 370,791 hectares the company received in compensation for its survey of Cantón Guerrero to Luis Riba and his wife María, who in 1909 sold it to the North Western Railway Company, a Canadian firm.[7] By 1905 W. M. Humphrey, a U.S. citizen, owned land in Cantón Bravos that had been given to the company.[8]

Close to 1 million hectares retained by the company as collateral for a loan were later turned over to Enrique Creel in a court judgment. The company had borrowed $200,000 from Manuel Angel Velasco in 1890, securing the loan with fourteen parcels of land in Chihuahua totaling 1,053,788 hectares, all of which originated in Valenzuela's survey concessions. All the mortgaged land was held by the two Valenzuela brothers and Celso González, except for six lots in former cantón Aldama that were owned by the Valenzuelas, González, and Herrera. Enrique Creel acquired the $200,000 note in 1901 and demanded payment. Six years of litigation followed, and the debtors' estates were still unable to pay off the note. As a result, the following lots, totaling 950,255 hectares, were turned over to Creel in 1908: 334,594 hectares in former cantón Meoqui (then Camargo); 61,246 hectares and 48,606 hectares in Camargo; 169,520 hectares in Iturbide; 178,275 hectares in Jiménez; 67,672, 36,892, and 13,040 hectares in Bravos; 40,480 hectares in Rayón (zone A of former cantón Aldama).[9]

Less than two months after the Gómez-Guerrero company received title to 1,007,666 hectares in compensation for its survey in Cantón Galeana, it sold Asúnsolo 136,968 hectares of the land at 28½ centavos per hectare, which was 8½ centavos above the government-set price of baldío in the state. The land was contiguous to another lot of unspecified size that Asúnsolo received as the return on his investment in the concession.[10]

Much of the land surveyed by Rafael García Martínez in Durango was quickly dispersed among numerous partners in the company and outside buyers. Some was even set aside to pay bribes. Eleven of the sixteen titles given for land in compensation were, at the company's request, issued directly to third parties to whom the company had already promised land.

The charter of La Esperanza, the company organized by García Martínez, Carlos Ruiz, and Celestino de la Quintana, specified that 10 percent of the terrenos baldíos the company would receive in compensation under the survey contract would be separated for the people "who may have lent services and assistance in order to obtain the concession," and 90 percent would be divided into ten shares, distributed in the following proportion: five for García Martínez (the managing partner and administrator), three for Ruiz, and two for Quintana.[11]

By 1890 the 10 percent had been distributed before a Durango notary. Ruiz, García Martínez, and the estate of Quintana (who had not lived to enjoy the fruits of the contract) then agreed to dissolve La Esperanza and to distribute their own shares in the agreed-upon manner. In 1890 the company held 278,006 hectares in Mezquital and 5,119 hectares in San Dimas, in land given in compensation for the surveys, which had been titled either to García Martínez or the company.

The Mezquital lands were divided as follows:

1. To García Martínez, for paying debts owed by the company, 81,636 hectares; and in payment of his five shares, 88,062 (total: 169,698 hectares)
2. To the (anonymous) individuals who helped the company get the survey contract, 17,612 hectares
3. To Ruiz for his three shares, 53,715 hectares

4. To Ing. Enrique Sardaneta for his services, deducted from Ruiz' share, 1,756 hectares
5. To the heirs of Quintana for his two shares, 35,225 hectares

The San Dimas lands were divided as follows:

1. To García Martínez for his five shares, 2,326 hectares
2. To Ruiz for his three shares, 1,396 hectares
3. To Quintana's heirs, for his two shares, 931 hectares
4. To those who helped get the contract, 465 hectares[12]

A year later the new firm of Rafael García Martínez y Socios sold its right to survey in the municipalidad of Villa Ocampo to Guillermo Beckmann, a German citizen. Beckmann paid the company $3,000 for the land and agreed to pay the costs of the survey. He received title to nearly 60,000 hectares in 1895. By 1908 the government still owed him 4,441 hectares for the survey, but Beckmann had by then sold his rights to the remainder for $1,500 to Saturnino Martínez, who in turn sold part of his right to it to Narciso Talamantes.[13]

Six of the seven titles given for land surveyed by Joaquín Casasús in Durango were issued directly to third parties. There is no evidence that they were owners of the company; Casasús merely requested that the titles be issued to "diverse individuals, with whom private arrangements have been made."[14]

After Macmanus e Hijos surveyed in Sonora under the Ornelas contract, they bought the two-thirds they had surveyed and then sold all 60,853 hectares to two foreigners, John S. Robinson and Frank W. Witbeck. By 1887, the Macmanuses told Fomento, the men had populated "a desert, inhabited only by the savage," with "a great number of families" and more than five thousand head of cattle. In 1901 all the land was resold to the Turkey Track Cattle Company.[15]

Another Sonoran surveyor, General Francisco Olivares, ceded to his foreign partner, John C. Beatty, his title to one-third of the 3,006,370 hectares they surveyed in the district of Altar (i.e., 1,002,123 hectares) in 1890, a month after the title was issued.[16]

In 1894 Manuel Peniche's company transferred three of its lots in Sonora, totaling 651,034 hectares, to the Wheeler Land Company, whose owners were George F. Wheeler, Frank M. Watts, Antonio Balderrama, and T. López del Castillo. The company still held the property in 1909.[17]

The 78,332 hectares that Lauro Carrillo earned for his survey in Sonora were ceded to Telésforo García in 1899 to repay a $3,000 loan.[18]

When the Mexican Land Colonization Company decided to concentrate its "resources, its personnel and energies" in Chiapas,[19] it arranged a land swap with the Mexican government in 1904 that resulted in the return to Fomento of twenty-two lots that had originally been titled to the company for surveys in Sonora. The twenty-two lots, scattered in the Magdalena, Guaymas, Alamos, and Altar districts, totaled 854,592 hectares. The company also returned a 66,160-hectare lot in Tonalá, Chiapas. In return, the company was given 171,407 hectares in Tonalá, another lot of 33,080 in the same district, and 32,209 hectares in Pichucalco, Chiapas. The deal also extinguished a credit of $64,243 owed to the company by Fomento. In addition, the company was to receive two-thirds of all the land it surveyed in the Chiapas districts of Tuxtla, Chiapa, La Libertad, and Comitán. The latter quantity was unspecified in the document memorializing the transfer, probably because the surveys were so bad and were being redone (see chapter 3).[20]

A letter from company lawyer Emilio Velasco indicated that the two-thirds of land in the four districts totaled 665,436 hectares, a figure Fomento was evidently reluctant to accept. Assuming the amount was roughly accurate, the company gave the government 854,592 hectares in Sonora and 66,160 in Chiapas (plus the $64,243 credit it was due) for about 900,000 hectares in Chiapas.[21] The company was obviously eager to cash in on a coffee-driven land boom in the southern state. As early as 1891 a company agent reported that the firm was busy subdividing and selling part of Soconusco for coffee cultivation, at prices he insisted were not "exorbitantes," but for a mere $12 per hectare. Lowland properties unsuited for coffee were going for $6 per hectare, and anyone could pay in installments.[22] By 1900 the company was complaining that it was having to delay land sales because Fomento still had not issued the title it was due for land in Tuxtla; the company was allowing prospective buyers to occupy the land they wanted and promising them it would be sold to them once the company received title.[23]

Overall, the proportion of land sold by the company was quite small, at least until 1905. In that year the company sold 51,755 hectares of the 317,875 it had been titled in Soconusco; 1,860 of the 246,748 it had received in Pichucalco; and 8,060 of the 389,705 it had received in Tuxtla and Chiapa. Of the 1.79 million hectares it had been titled in Chiapas, it had sold the rather insignificant quantity of 61,665 hectares, or just 3 percent, by 1905. The balance still held of 1,730,030 hectares was sold in 1906 to the Land Company of Chiapas for 170,000 pounds sterling plus stock in the Land Company.[24]

But between 1909 and 1916 in Soconusco, the firm sold 173 lots totaling 18,764 hectares to Mexicans, and 35 lots totaling 15,807 hectares to foreigners. In Motozintla in the same period, 6 lots totaling 5,044 hectares went to Mexicans, and 10 lots totaling 8,228 hectares went to foreigners. An additional 64 lots totaling 12,953 hectares were "promised" to Mexicans (50 lots, 8,739 hectares) and foreigners (14 lots, 4,214 hectares) in Soconusco, and 16 lots totaling 3,502 hectares were promised to Mexicans only in Motozintla.[25]

The company continued to maintain "Margaritas," a 5,000-hectare dyewood finca in Tonalá, and a 20,000-tree coconut plantation in the same department. It also had a timber operation and a coffee and hule plantation in Soconusco. A "número considerable" of smallholding tenants worked company lands in various departments. All of the company's own operations were still underway as late as June 1917, but the record indicates that they were being abandoned by September.[26]

When Luis Martínez de Castro was authorized to survey in Chiapas in 1901, he agreed to share with his chief surveyor and legal representative, José Tamborrel, a vecino of Tenosique, Tabasco, some of the land that would be granted in compensation. In return for his services, Tamborrel would receive two-thirds of the land that Martínez de Castro would be given in compensation, and the latter would also assign Tamborrel his right to buy the government's two-thirds. Moreover, in any composiciones arranged under the contract, Tamborrel would get 60 percent of the net profits.

By 1906 Martínez de Castro and Tamborrel jointly held a 107,854-hectare lot, and another one of 12,380 hectares in Chilón department; one-third of the total belonged to Martínez and the balance to Tamborrel. The latter resold 40,000 hectares to three men for $1.20 per hectare, as well as his rights to an adjoining 80,000 hectares (whose title he had not yet received) for $0.10 per hectare. A few weeks later Tamborrel and the three men formed a joint-stock company for the "explotación de terrenos" in Chilón. In a separate transaction, Martínez de Castro and Tamborrel sold 10,000 hectares to Bulnes Hermanos, the Tabasco surveyor, for $2,500 and survey data that Bulnes had collected in its own survey of Chiapas. In December 1906, Martínez de Castro sold what was left of his share in Chilón (estimated at between 25,000 and 29,000 hectares) to Eduardo Hartmann for $25,000. Further subdivisions and sales were recorded between 1906 and 1909, in 1937, and in 1953 for the lots originally held by Martínez and Tamborrel.[27]

A lot surveyed in Sinaloa by Mariano Gallegos in 1886 was subdivided and resold in seven different transactions before 1902.[28] The one-third received by Becerra's company for surveying "Medano del Pozoli" was sold to the Compañía Explotadora de las Aguas del Río Fuerte, S.A., which also purchased the government's two-thirds. The company still held the land as late as 1921.[29]

LAND PRICES

Because a survey increased the value of land, its effect on land prices was invariably positive. This section will estimate the proportional increase in land value as the result of a survey. In most cases, the value of newly surveyed land is compared to the government-administered price for the land (terreno baldío) before it was surveyed.[30] Thus the price of unsurveyed land is not a true market price, but because unsurveyed land could not be sold except by the federal government and for the price that it alone fixed, no other price is relevant.[31]

When Fomento received an offer to buy recently surveyed terreno nacional in Durango, one official argued that because the land was surveyed, it should be sold for 50 percent more than the rate set for unsurveyed terreno baldío.[32] Luis Martínez de Castro's offer to pay the legislated baldío price of $1.10 per hectare for already-surveyed land in Sinaloa was rejected in 1893 because "the survey has already cost the government." Fomento offered it to him for $2 per hectare instead, payable in public debt instruments, or 81 percent over the baldío price.[33] In three other cases in which Fomento sold recently surveyed land in Sonora, Sinaloa, and Chiapas between 1888 and 1900, the price varied between 50 percent and 81 percent of the price set for unsurveyed land in the corresponding years.[34] In Sonora in 1903, unsurveyed baldío cost $1.10 per hectare, while surveyed terreno nacional cost about 63 percent more, or around $1.80 per hectare. In the same year in Chiapas, baldío went for $3 per hectare and terreno nacional for 66 percent more, or about $5 per hectare.[35]

After the Sinaloa Land Company's contract expired in February 1908, it no longer enjoyed the privilege of buying the government's two-thirds at whatever price the government had fixed for unsurveyed baldíos—$1.20 per hectare from 1901 to 1907. After baldío prices in that state had jumped to $4 per hectare in 1907, Fomento offered surveyed terreno nacional to the company for $6 per hectare in bonds, or 50 percent over the unsurveyed price.[36] Other buyers were offered surveyed land for the same price in 1907.[37]

As noted in chapter 3, a Fomento ministry official estimated that the value of Chiapas land surveyed by the Mexican Land Colonization Company increased some 73 percent, apparently only by virtue of the survey.

CONCLUSIONS

In all but two cases in which evidence appeared that a survey company transferred all or part of the land it received in compensation for a survey, the land transferred was subdivided at least once.[38] The two exceptions were the Macmanuses' sale of all their land to two Americans, and Lauro Carrillo's transfer of all his to Telésforo García to satisfy a debt. The Mexican Land Colonization Company's holdings in Chiapas were only marginally affected by transfers. It was the only survey company for which evidence was found of an effort to permanently maintain possession of the land it acquired. As noted earlier, the company exploited the land for its coffee, hule, and timber.[39]

The methodology of this chapter, largely dictated by the nature of the evidence in the Archivo de Terrenos Nacionales, is too limited to submit a quantitative test of the hypothesis that the use of the survey companies increased the concentration of land in Mexico. The data, however, are exhaustive for all cases involving transfers subsequent to surveys, and they do indicate that land transferred by the companies was usually subdivided. It was rarely sold *in toto*.

Of all the contractors who eventually received land in compensation for a survey (see Table 9), enough were found to have subdivided and transferred their land to indicate a pattern of subdivision and transfer. Further evidence pointing to a dilution of landownership was provided in chapter 2, which indicated that most of the companies were formed by partners who commonly distributed the land grants among themselves. This distribution often constituted the initial subdivision of the land, which was further subdivided when it was sold to outsiders or lost to creditors.

This pattern is inconsistent with the argument that the companies' land grants resulted in greater land concentration. It is nevertheless possible that because the companies received such large properties, the average size of all landholdings increased as a result of the government's survey policy. If the companies obtained and kept very large lots, overall distribution would have been affected, and big landowners would have made up a larger proportion of all property holders. This sort of test, however, was not only beyond the scope of this project's resources but, as noted at the beginning of the chapter, probably beyond the limits of verification.

This chapter also presented evidence of changes in land values as a result of the surveys. We have already seen that one of the reasons the state undertook surveys of the public lands was to reduce the risks of ownership. Secure titles with clearly defined boundaries were necessary to attract the capital investment in the country's natural resources that the Díaz government sought. Without surveys and the enforceable system of property rights within which the surveys were embedded, the transaction costs faced by risk-taking entrepreneurs to enforce their property claims would have been prohibitive. The data presented in this section indicate that the government's survey policy added at least 50 percent to the price of its public lands, a rough measure of the value of the reduction in transaction costs. The rise in land values thus indicates that the surveys generally benefited landowners, whose possession of land became more secure as a result of the surveys.

If landownership in Díaz-era Mexico did not become more concentrated as a result of what the survey companies did with their grants, and if the benefits of better-defined property limits extended to smallholders, the argument that the survey contractors contributed to the outbreak of the 1910 revolution is much less tenable. That most of the surveys were completed by the early 1890s, nearly two decades before the outbreak of the revolution, likewise casts doubt on the argument. The possibility that a significant number of public land squatters were deprived of their holdings by the surveys, and that the resulting increase in value of the newly surveyed land thus placed it beyond their reach, has already been disposed of in chapter 4.

The next chapter will explore the new revolutionary governments' treatment of the survey company land grants, and will offer a final assessment of the companies' contribution to the discontent that led to the revolution.

6 The Survey Companies and the Revolution of 1910

In 1902, a little more than eight years before his resignation as the president of Mexico, President Díaz banned the government from hiring private companies to survey the public lands. The executive branch, according to the new law, would be empowered to conduct surveys of baldíos by means of *comisiones oficiales* (official commissions); surveys by private concerns were expressly forbidden, as was the practice of compensating them with land.

The secretary of Fomento explained that although the practice of hiring private contractors and compensating them with land they surveyed had enabled the state to take possession of much of the country's unoccupied land, and had also encouraged the perfection of titles and amortized a sizable portion of the public debt, the survey companies "inconvenienced, on many occasions, the holders of rural property, and on their behalf the law [of 1902] contains very clear and precise precepts guaranteeing their rights."[1] A longtime critic of the survey companies, Fomento official José Covarrubias, believed the decree would bring more tranquility to the countryside. He asserted that declaring "possessed lands" no longer to be considered part of the national domain would prevent any further acquisition of worked land by speculators. In the future, speculators would be confined to buying land on which "there is absolutely no possessor nor any created interest."[2]

Yet the repeal seemed almost irrelevant. Only three authorizations for companies to survey had been issued in the preceding ten years, and roughly 87 percent of all the land that the companies would ever receive in compensation had already been given away more than nine years earlier. The president was outlawing a practice that had been virtually abandoned for nearly a decade. No clear reasons for the timing of the decree are now evident, except that it was one of a series of reforms aimed at protecting the rights of individuals who occupied land without holding title to it.[3]

Díaz was finally driven from the presidency in May 1911 by scattered guerrilla insurgent units nominally headed by Francisco Madero. One of the most durable components of the traditional interpretation of the Mexican Revolution of 1910 has been the emphasis on the Díaz government's public land policy as a source of revolutionary discontent, a topic that we began to explore in chapters 2 and 3. Four tasks await our attention here. First, we will review and weigh the meaning of the origins of that interpretation. Second, the extent to which the companies contributed to the grievances that drove the revolution will be considered, partly by reference to recent research into rural unrest during the Porfiriato. Then an overall assessment of the quality of the surveys will be attempted, in part by reference to contemporary evaluations. Finally, the treatment that the victorious revolutionaries accorded the survey companies and the property they received in compensation will be reviewed. All the revolutionary factions, including that which finally triumphed in 1917, to some degree opposed the public land policy of the ancien régime. How was this opposition actually expressed when the smoke cleared and the new regime moved to consolidate its authority?

HISTORY AND THE COMPAÑÍAS DESLINDADORAS

The historiographical tradition of attributing to the survey companies so much of the rural discontent that sparked the revolution of 1910 was rooted in a single book, *Legislación y jurisprudencia sobre terrenos baldíos*, by Luis Wistano Orozco. It was published in 1895 at the zenith of the regime's power and prestige. Its author was the most influential of the few dissident voices heard in Porfirian Mexico before 1908, and provided future critics of the regime "an intellectually unimpeachable indictment of the system."[4] Orozco's attack on the compañías deslindadoras was actually part of a powerfully worded condemnation of the regime's public land policy. His purpose was to promote a reform that would distribute land "among the greatest possible number of men," because "more proprietors means more happy men." His vision of an ideal society was one where "the contrast of proud opulence and abject poverty" does not exist but where "the orderly division of wealth guarantees everyone's dignity and independence." One of the best means available to uplift the "disinherited" was to divide the public lands among them.[5]

"Las Compañías Deslindadoras," Orozco argued, "have sometimes speculated and been lucky, and other times have lost their money; but in no case have their labors benefited either agrarian property or the country's colonization."[6] The surveys failed not only to break up the gigantic haciendas that Orozco blamed for Mexico's backwardness. They also resulted in the usurpation of land from poor peasants, because

> it is not the powerful, it is not the big hacendados who have seen those millions of hectares fall from their hands, but the poor, the ignorant, the weak, those who are not able to call a District Judge, a Governor or a Minister of State "compadre" [lit. godfather, often pal]. Hence a complete perversion of the purpose of the laws and of the ideals of democracy; since while the supreme purpose of the laws regulating baldíos and colonization is to extend the benefits of agrarian property to those who lack it, in the name of those very laws poor peasants are stripped of their possessions, or they are obliged to ransom them by means of painful sacrifices.[7]

Smallholders, unable to afford a lawyer who could oppose the survey in court, were forced into a deal with the "shameless thieves who call themselves surveyors," paying off the survey company to protect their land. The companies, which had ruined some speculators while enriching others, had to be suppressed "forever," Orozco concluded.[8]

To Orozco, the survey companies alone were not to blame—the entire public land policy was a failure because it put smallholders at the mercy of anyone who entered a claim for public land. When a claimant had a reputed baldío surveyed, which he had to do under the land law of 1863, anyone occupying the land had to prove legal possession, if not ownership, or risk eviction. The public land laws were not in themselves the problem; the evils done in their name were "economic errors on the practical business level. Everything will therefore be reduced to rectifying the means now employed in affairs involving the baldíos."[9] Consequently, Orozco proposed two reforms: Remove from the present law "every possibility of attacking small properties" (i.e., less than one sitio, or 1,756 hectares), and simplify the claims process.[10]

Orozco's denouncement of the survey companies was not without foundation, of course. A few of the companies had acquired immense shares of territory, generating numerous complaints of usurpation in doing so. The foreign ownership of Luis Huller's Baja California concession in particular had inspired resentment among members of the political elite, who also had good cause to suspect the company of plotting to annex Baja California to the United States.[11] But as a source for the public land history of Mexico, Orozco's book is best understood as the expression of the views of the agrarian-oriented and romantically inclined intellectual community that envisioned a radically different path to economic development than that of industrial capitalism. The author's ideas were part of a tradition of nineteenth-century populism that Gavin Kitching has traced to Sismondi, Proudhon, and the Ricardian socialists. Like them, Orozco was an urban intellectual whose opposition to industrialization lay in his conviction that it tended to aggravate inequalities that could only be overcome by a system of small-scale enterprise.[12] Orozco's commitment to this vision of the future inspired his powerful critique of Porfirismo. While magisterial in its historical treatment of Mexico's public land laws, the book's sweeping judgments of the regime's land policy (which make up less than one-tenth of its volume) are less the work of a scholar than that of an advocate. This is said not to diminish the book's power or importance, but simply to challenge its authority as the foundation for an assessment of the government's land policies.

Orozco's arguments would subsequently become part of what D. A. Brading called the "populist tradition" of revolutionary historiography, an interpretation of the revolution as a peasant uprising to recover land stolen from the peasants by hacendados who monopolized huge tracts of underexploited real estate.[13] This tradition has been sharply revised in recent years, so much so that Brading wondered whether the peasantry was not so much a part of the revolution as its victim.[14] In any event, Orozco's proposal to increase the number of smallholders at the expense of the big landowners was adopted by Andrés Molina Enríquez, who quoted extensively from Orozco in his *Los grandes problemas nacionales,* easily the most influential of the critical works that began to appear late in the Díaz era as the revolution was about to unfold.[15] The principal anglophone works in this tradition were produced within the first twenty-five years of the revolution. The compañías deslindadoras, Frank Tannenbaum wrote, "led to a rapid transfer to a few favorites of vast areas of government lands," and "a systematic attack . . . against the existing occupants of the land."[16] Helen Phipps claimed that "entire villages were taken over by the companies," while George McC. McBride limited his criticism to stating that most of the land granted the companies "was held in immense tracts for speculative purposes."[17]

All of the authors in the populist tradition shared the belief that what Mexico needed in 1910—and what the revolution would provide it—was the destruction of large landed estates and their replacement with small farms. Most of these writers, like Orozco, believed above all in the superiority of property ownership, which Orozco called "a public necessity," a right originating in the "rational essence of man, who . . . is properly and truly the sovereign owner of the world."[18] They were also like Orozco in another way. When they condemned the survey companies they invariably failed to produce supporting evidence. Orozco's judgment alone appeared to suffice.

The populist argument was repeated widely after the revolution, in part because it provided further justification for the cause of the revolutionary faction that finally triumphed in 1919. The historiography that followed the revolution was profoundly shaped by the revolutionary state itself, which felt it necessary to "sanction with the past the political order" that it established, as Enrique Florescano has pointed out. What had been a government of "paz y progreso" turned into "La Dictadura," while "the large estates, the big landowner and the exploitation of the Indian take the place which in liberal works had been occupied by the might of the Church, the Spaniards, the priests and religious obscurantism. A great portion of the country's long past was satanized in order to justify the social and political order that the Revolution sought to create."[19]

THE SURVEYS AND THE COMING OF THE REVOLUTION

Chapter 4 presented evidence of grievances against the companies. Is it possible that some of those grievances evolved into revolutionary violence after Francisco Madero's call to arms in November 1910? An answer to this question would be the most satisfactory way of resolving the issue of the companies' contribution to the revolution. Unfortunately, the question is answerable only in the most speculative way. First there is the problem of causation. Revolutionary outbreaks after November 1910 could be attributed to numerous causes; it would be necessary to isolate survey company activity as an independent variable, match it with a specific outbreak of violence, and weigh the companies' contribution to that violence. One particularly thorny question is how to make the match. If the variables are matched purely on geographical terms (i.e., village X revolted, village X was surveyed by company Y), can we be sure that the mere presence of the company at a certain time inspired the revolt? Perhaps a column of revolutionaries simply rode into village X, collected a few recruits, and stood off the *federales* (federalists) before marching on. Did the recruits join because they or their fathers lost land to the survey companies? How much time would have to elapse before one would have to abandon the company's activities as a causative factor? The documentation required for such a finding would need to be prodigious.

In the absence of the resources necessary for an archival investigation of this issue, recent secondary literature dealing with the connections among land usurpation, rural unrest during the Porfiriato, and the outbreak of the revolution in various regions of Mexico was searched. The survey companies were mentioned as causative factors in four works of the fifteen selected for this study.[20]

Paul Friedrich's classic *Agrarian Revolt in a Mexican Village* reported that the land owned by the Michoacán village of Naranja was practically stolen by a survey company in the 1880s, an event that had "catastrophic" consequences for the local economy. However, little effort to recover the land was made before the 1920s, when the appropriate leadership finally developed.[21] Another work reported that twenty-nine years before the Huastecan Indians of San Luis Potosí revolted in 1910, the government had surveyed their community lands and those of wealthy landowners, offending both groups. When the Indians revolted, they were fed up with the "continual loss of their lands and the ineffectiveness of their claims," and feared a newly scheduled cadastral survey.[22] In the third case, the identification of interests common to *finqueros* (farmers) and Indians in Chiapas, inspired by the appearance of the survey companies, led them into an alliance against Díaz.[23] Finally, although land thefts by surveyors in Chihuahua have been identified, no evidence of protests by the inhabitants has been provided, except for one village, Tomochí. Other historians attribute Tomochí's revolt to different factors, however.[24]

In each of the four cases, some association is made between the survey companies and revolt. But the connections are frustratingly vague, particularly in terms of the timing of revolt in relation to the surveys. In the other eleven studies cited, land usurpation resulted from the enforcement of Article 27 of the 1857 constitution (mandating the division of ejidos) in two cases (Buve and Womack), and in another case from public land claims by a wealthy family (Joseph). In other cases—notably those reported by Schryer, Jacobs, Menegus Bornemann, and Berry—federal land policy was carried out peacefully and was not connected with any uprisings.

Peasants undoubtedly lost land during the Díaz era and joined the revolution as a result. But today we know that peasant grievances over land were only a contributing factor to the longest and bloodiest revolution in twentieth-century Latin America. The source of those grievances undoubtedly lay in a combination of federal land policy (the division of ejidos, public land claims, survey company abuses) and expropriations by individual landowners. Such grievances may well have been pending for many decades before 1910, when conditions for revolt appeared most propitious. It is also possible, as suggested by the evidence in chapter 4, that the federal government may have shielded peasants from land grabbers until sometime after 1900, when the regime gradually lost control over its political allies in the states. Friedrich Katz attributes Díaz's downfall in part to the president's inability, in the last decade of his rule, to continue the policy of divide and rule among the regional elites; a "científico offensive" inspired the opposition of peasants as well as merchants and hacendados who were excluded from power.[25]

ASSESSING THE RELIABILITY OF THE SURVEYS

Before reviewing the response of the revolutionary regime to the Díaz government's survey policy, some conclusions about the quality of the surveys will be drawn. The assessment will be based on the archival evidence regarding the accuracy of the surveys as presented in chapters 2 and 4, and on some general evaluations made by the government itself.

Although the surveys were still consulted to settle property disputes many decades later, their reliability never ceased to be questioned.[26] President Díaz's 1902 decree forbidding the hiring of private companies to survey the public lands was officially attributed to the annoyance these companies caused to individuals who were occupying rural land, whose rights the 1902 decree was aimed at protecting, as noted earlier. But doubts about the quality of the surveys probably inspired the decree as well. When the vecinos of Villa Ocampo, Durango, requested copies of Rafael García Martínez's survey documents in 1903, a Fomento official noted in a memorandum that the request should be denied "because of the slight confidence that one must place both in this survey and in all the surveys in general."[27]

One of the most penetrating critiques of the surveys originated within Fomento itself in 1903 but was not publicly available until after the revolution, when President Madero's Fomento secretariat published it. The essay was written by José Covarrubias, the head of the Fomento section in charge of the terrenos baldíos, and an adviser to Fomento chief Leandro Fernández on the drafting of the decree of December 30, 1902, that abolished the use of the companies. Land acquired under the law of 1883 was supposed to be used for colonization, but the poor quality of the land and the inaccuracy of the maps, Covarrubias wrote, made colonization impractical.[28] When the surveys authorized by the 1883 law were undertaken, he pointed out, the government lacked the funds to survey the land itself. Thinking that the great benefits of a survey would be obtained without cost, it failed to consider one of the main causes of the "disappointments" that surfaced as a result of the surveys. Few physical traces of the surveys remained—stakes disappeared and vegetation covered openings in the ground. But for Fomento to have sent an expert to check surveys that the contractors had already conducted would have duplicated the cost of the original operations. And because the government's objective was to avoid spending any money at all on surveys, such inspections were out of the question. As a result, "many" survey companies submitted maps that were inaccurate, providing only "approximate data" often gathered without even measuring the land. Not only were the borders of the baldío wrong, but titled land was often included as baldío because the owners had not been notified to present their titles. For these reasons, Covarrubias pointed out, the 1902 decree banned any further use of survey companies, and official survey commissions had to be substituted.[29]

The Díaz regime's last official word on the subject of the reliability of the surveys was embodied in a December 1909 decree suspending the sale of all terreno nacional until it was resurveyed by official commissions. Regulations issued six months later established a Dirección Agraria within Fomento that would carry out the resurveys; it stipulated in greater detail than ever before the technical and legal procedures that the survey parties were to follow. Fomento chief Olegario Molina attributed the legislation to the "vagueness and imprecision" of the government's knowledge of the public domain, as well as to "the difficulties that for that reason used to take place in transactions involving that category of property." The government survey parties went to work first in Chihuahua's Bravos district, along the U.S. border, and among the pueblos between the Yaqui and Mayo rivers in Sonora, but revolutionary violence forced them to suspend Porfirio Díaz' last attempt to map the public lands.[30]

The reliability of the surveys that the Díaz and González governments contracted for was probably commensurate with the amount that the government was able to pay for them—that is, as little as it possibly could, and then only in the form of in-kind compensation of land about which it knew little or nothing. Considering the penury of the state in the 1880s and the intense pressure that existed to discover, map, and perfect titles to public land as quickly as possible, the surveys it got were probably the best the state could afford under the circumstances; while often crude and occasionally fraudulent, the surveys' contribution to the surge in economic growth that characterized the Díaz era is undeniable. The natural barriers of Mexico's topography undoubtedly presented the greatest challenge of all to the survey technology of the day. As we saw in chapters 2, the least reliable surveys were those conducted in the two southern states of Tabasco and Chiapas, where the dense jungle terrain made accurate measurements and the establishment of permanent markers almost impossible; the hazards confronted in the mountains and deserts of northern Mexico were considerably milder in comparison.

THE REVOLUTIONARY GOVERNMENTS AND THE COMPANIES

Francisco Madero's Plan de San Luis Potosí, issued in October 1910, was the revolution's initial call to arms. The document contained a single paragraph that spoke to the agrarian question and the regime's public land policy, which it condemned for usurping peasant lands: "In an abuse of the baldíos law, many smallholders, most of them Indians, have been stripped of their lands by decree of the Secretaría de Fomento or by court judgements." The future president pledged the restitution of such lands to their former occupants, plus the payment to the occupants of an indemnity by those who acquired the lands.[31]

Five years later, when the Constitutionalist faction was fighting for control of Mexico against the more populist and agrarian-oriented forces of Pancho Villa and Emiliano Zapata, the Constitutionalist chief, Venustiano Carranza, promulgated legislation directed entirely at restoring to communities lands they had lost since the reform law of 1856. It condemned not only the abuse of the 1856 law, which instead of dividing community land among the inhabitants merely eased its transfer to "a few speculators," but also the plunder of all sorts of commonly held land by means of "concessions, composiciones, or sales arranged with the ministries of Fomento and Hacienda, or under the pretext of measurements and surveys, in order to benefit those who submitted claims to excedencias or demasías, and the so-called survey companies." Declaring that the return of stolen land to the communities was "un acto de elemental justicia" and the only way to ensure peace and to promote the general welfare, Carranza ordered his generals to expropriate land and give "sufficient land to the towns that lack them." The decree stressed, however, that such land was not to be given in common, but divided in fee simple. Furthermore, all land entitlements or surveys by the federal government since December 1, 1876—when Porfirio Díaz first took power—that resulted in the invasion or usurpation of any community land were nullified.[32]

Carranza's 1915 land reform law was incorporated in the federal constitution of 1917, whose Article 27 nullified all official acts since 1856 that may have deprived communities of land, and further required the restitution of such land. Separately, the article empowered the state to provide communities lacking sufficient land or water by taking those resources away from adjoining private properties. It ordered Congress and the state legislatures to fix maximum limits on the size of individual estates, whose excess over the maximum was to be subdivided and sold by the owner. Of principal interest is the last paragraph of the article, which provided the basis for the new regime's action against the landholdings of the survey companies: "All contracts and concessions made by previous governments since the year 1876, that may have resulted in the monopolization of lands, waters, and the natural riches of the country by a single person or society, are declared subject to review, and the Executive of the Union is empowered to declare them null when they imply grave damage to the public interest."[33]

The way was opened, therefore, for the nullification by presidential fiat of any survey contracts—and thus of titles to land granted to the companies in compensation for their labors—that damaged the "interés público."[34] The rights of third parties that subsequently acquired land from the survey company were usually recognized, unless the land was directly assigned by the company and Fomento had issued the title to the third party at the company's request.

Ignacio Sandoval's survey of Cantón Mina in Chihuahua will serve as an example of the revocation process. The survey was reviewed in 1919 by Gabriel Moreno, an official in the Dirección Agraria, who concluded that the survey was illegal on the following grounds: The surveyor failed to record longitudes and the size of angles, thus violating established procedure; no markers were constructed; the survey was never approved by Fomento or even checked by it; the land was not appraised; the surveyor presented no credentials, and because he was a partner in the company his work could not be considered impartial. Furthermore, the 1883 law under which the survey was conducted was intended to promote colonization, but the company showed no interest in colonizing the land it received. Although its contract limited the company to survey "baldíos colonizables," it surveyed the whole cantón instead, most of which was unfit for colonization. Finally, the granting to the company of 1.86 million hectares, covered with "construction-grade timber" constituted "an extremely gross violation of the law with grave damage to the Assets of the Nation." Moreno, citing the last paragraph of Article 27 of the constitution (as quoted earlier), recommended the nullification of Sandoval's contracts and titles.

Moreno's recommendation exaggerated Sandoval's land grant by a factor of three, confusing the total surveyed with the quantity actually titled to the company. His argument was challenged by the department's lawyer, Ramón de la Barrera, who asserted that neither the titles nor the contracts could be nullified because they were provided and conducted legally under the 1883 law. Sandoval's legal possession of the land had never been successfully challenged, the lawyer continued, adding that

Article 27 did not apply because there was no *acaparamiento* (monopolization), nor was the public interest damaged. Finally, de la Barrera argued, the survey, "far from causing damage to the public interest," gave the nation the benefit of two parcels of land equal in size to that given Sandoval.

Moreno's view was upheld by his superiors, and Sandoval was notified in December 1923 that his titles to the 633,537 hectares he received would be nullified for two reasons: His contract and the titles were issued "under the condition" of colonization, and his possession of the 633,537 hectares was a "notorious monopolization of land" in violation of Article 27.[35]

Sandoval was given thirty days to submit a defense. President Obregón nullified the concession on November 21, 1924, nine days before he yielded the office to President Calles. The official decree repeated Moreno's criticisms, and concluded that the land grant was "a notorious monopolization of land and natural riches of the Nation by a single person, with grave damage to the public interest." Article 27 was cited as the source of authority.[36]

Because Obregón's decree also nullified the titles of anyone who had subsequently acquired any of the land from Sandoval, these third parties were given sixty days to petition the government for recognition of their titles, a process that some had already initiated. One of the parties, the Pine King Land & Lumber Company, had filed a claim for the land before the joint U.S.–Mexican Claims Commission.[37] Not until 1930 did the Mexican government settle with the company. The government recognized that Pine King bought 209,111 hectares in good faith from Luis García Teruel (who had acquired it from Sandoval). Pine King in turn agreed to survey, subdivide, and sell all the land within three years, preferably to Mexicans living on the land, at prices to be approved by the government.[38] The government's concern that small farmers who acquired land out of the Sandoval concession may not have seen the legal notices in the official publications prompted a 1928 decree giving anyone with holdings under 5,000 hectares six more months to legalize their possession; a similar decree was issued in 1930.[39]

The effort to clarify property rights in the Sandoval concession continued into the 1970s. In response to a request for information to resolve disputes over property lines that came from the Confederación Nacional Campesina's Comité Regional Campesino in Guadalupe y Calvo, Chihuahua, the government said in 1958 that it still was not sure how much terreno nacional was left in the Sandoval concession. A survey, the official wrote, would be necessary. As late as 1972 the Dirección de Terrenos Nacionales recommended that in accord with the provision of Article 27 authorizing the nullification of contracts and concessions made after 1876 (as quoted earlier), the present occupancy of the Sandoval concession of 633,537 hectares be studied.[40] Thus some fifty years after Madero's call to arms, the government of Mexico, guided by the party of the Institutionalized Revolution, could still be found fussing with the transgressions of Porfirismo, revealing a mixture of ineptitude and ideological decay far beyond the capacity of its nineteenth-century predecessor.

The postrevolutionary state's handling of the Sandoval concession was characteristic of its treatment of those survey contractors whose rights it would challenge. (See Table 16.) Typically, Gabriel Moreno submitted a strong recommendation in favor of revocation within two years of the passage of the 1917 constitution. Many of the revocations were not announced until 1924, President Obregón's last year in office. Generous allowances were usually made for the recognition of rights to land acquired in the concessions by third parties, particularly if they happened to be smallholders. Those who claimed the land by virtue of the original title, such as a survey contractor's heirs or other legal successors, often tried to oppose the revocation by attacking the charge that the surveyor failed to colonize the land.[41] In this they were probably justified, because few of the contracts specified any obligations to colonize, notwithstanding the intentions of the 1883 law (see chapter 2). Many of the government's other charges against the companies were similarly inaccurate, vague, and unsubstantiated.

But the legal underpinnings of the revocations were scarcely relevant in view of the sweeping powers delegated to the president in Article 27 to expropriate property. Legalistic arguments seem to have been beside the point, as they usually are. Surely, no target for revolutionary agrarian justice looked safer than that handful of individuals who had torn through the countryside in the 1880s, collecting hundreds of thousands of hectares of vacant land under contracts authorized by the detested ancien régime. In the early 1920s, even the holdings of American citizens would have been less tempting because of the touchy state of U.S.–Mexican relations, which were severely strained by Washington's insistence on just compensation and by the ever-present fear of U.S. military intervention.

Most of the nullification decrees were based on little more than an inspection of the government's files, and the wording of the decrees was fairly similar. The exception was Policarpo Valenzuela's survey of Tabasco, which from 1912 was subjected to a detailed investigation by surveyors in the field who found numerous irregularities and inaccuracies. Valenzuela's operations "left much to be desired regarding their accuracy," according to one government surveyor. In 1912 an investigator

Table 16 Revocations of Survey Concessions after 1910

Date	Contractor	Area (ha)	Source
		Chihuahua	
11/21/24	Sandoval	633,537	D.O. 1/14/25, 254–56
6/26/24	Valenzuela	2,795,191	D.O. 8/14/24, 1831–33
6/26/24	Gómez del Campo	1,519,437	ATN. CHIH 8245
4/22/20	Sandoval y Pacheco		ATN. CHIH 78534
5/22/24	Ornelas	20,284[a]	D.O. 6/10/24, 624–26
		Durango	
1/8/25	García Martínez	671,414	D.O. 1/31/25, 639–42
1939	Calderón	95,679	D.O. 2/7/39, 3–5
3/21/25	Asúnsolo	347,699	D.O. 4/4/25, 1717–19
		Sonora	
4/3/24	Peniche	651,034[b]	D.O. 4/30/24, 1556–7
5/29/30	Olivares	1,762,379	D.O. 6/30/30, 3–4
		Tabasco	
1917	P. Valenzuela	—	ATN. TAB 1.71/8[c]

Note: D.O. is *Diario Oficial*.
[a]The decree only specified land that Ornelas was titled in Coahuila, but because it canceled all titles under his contracts, the land he received in Chihuahua alone is given here.
[b]Total not specified in decree.
[c]In letter from Fsco. L. Terminel to Sec. de Ag. y Fom., Aug. 29, 1930.

uncovered so many errors in the survey of the city of Teapa that he concluded that Valenzuela's surveyors had not even gone into the field. The firm's concession and titles were revoked in 1917, but litigation between Valenzuela's estate and the government over the decree continued into the 1930s.[42] A government official reported in 1935 that Valenzuela's estate still held more than 200,000 hectares.[43]

Revocations were not issued across the board, however, nor were they applied with an even hand. President Obregón's decree nullifying the concession of Ignacio Gómez del Campo and Ramón Guerrero was issued the same day as that canceling Jesús Valenzuela's, but its publication in the *Diario Oficial* was withheld "by superior decree" for at least four years. It could not be implemented until it was published.[44]

Some nullification decrees were not issued until much later, and other contractors escaped punishment completely. José María Calderón's title to 95,679 hectares in Durango was revoked in 1939.[45] The surveys by José María Becerra's company, conducted during the course of arranging com- posiciones, were actually praised by the Obregón administration, which said they were done "with complete regularity, and the titles were made with strict justice to various people on small parcels"; thus all the titles derived from the Becerra concession were "buenos."[46]

One of the biggest contractors to emerge unscathed from the revolu- tion was Luis Martínez de Castro in Sinaloa, and his successor, the U.S.-owned Sinaloa Land Company.[47] In 1912 President Madero signed two titles compensating Martínez de Castro for long-pending surveys in Sinaloa.[48] One survey by the Sinaloa Land Company, pending since before the revolution, was actually approved by the Obregón admin- istration.[49] During President Calles's term in office, a government sur- veyor was assigned to accompany a Sinaloa Land Company engineer as he located the company's one-third.[50] The revolutionary government even compensated Martínez de Castro for surveying 75,000 hectares in Chi- apas for which the Díaz government had refused to pay him. From 1905 to 1907 the Sinaloa Land Company's appeals for compensation were turned down by the Díaz government and the courts on the grounds that the survey had not been completed before the contract expired. In 1920 Martínez de Castro, as the representative of the company, again requested his one-third of the 75,000 hectares, and the following year the Obregón government allowed his widow to subdivide and sell 25,000 hectares in the department of Chilón as compensation. The subdivision was com- pleted and the titles were signed by President Calles in 1927.[51]

In a similar case in Chiapas, Romano y Compañía, a Spanish firm, had acquired permission from the Díaz government to survey 162,184 hect- ares, for which it was due 54,061 hectares in compensation.[52] The com- pany also bought part of the remaining two-thirds. President Carranza obliged the company with a title to 66,061 hectares in January 1920, de- spite the fact its owners were foreigners and the land was within one hun- dred kilometers of Guatemala, a circumstance prohibited by Article 27. President Calles nullified Carranza's title in 1925. President Emilio Portes Gil recognized the company's right to the land in 1929 on con- dition the firm subdivide and sell all of it, which was done before the 1944 deadline.[53]

The British-held Mexican Land Colonization Company and its succes- sor, the Land Company of Chiapas, also received special treatment. In re- turn for the company's renunciation of all claims to the 7.75 million hectares it was titled in Chiapas and Baja California (of which only 1.77 million hectares were in Chiapas), the government of President Obregón in January 1923 agreed to pay the companies $16 million.[54] The difference in treatment was partly because, unlike the land involved in most survey company contracts, much of that owned by the British company had been purchased by it for colonization purposes and had been improved at considerable expense.

The government's objective was to take back land given to the survey companies and their successors, not that sold by them to third parties, whose rights were usually upheld. The government in 1930, fearing that smallholders who acquired land in the nullified Gómez del Campo concession in Chihuahua, "through ignorance or obstacles in the means of communications," had failed to seek approval of their titles, gave them another six months. The rights of such "agricultures en pequeño" to the land they held were guaranteed. Others who owned land above the maximum set for their jurisdiction were given extra time to subdivide and sell it.[55] Deadlines were extended through 1938 to allow smallholders to legalize holdings obtained through the Land Company of Chiapas.[56] Even foreigners with sizable holdings were not automatically ejected. Still pending in the 1930s were the claims of Ivan Benton to 41,885 hectares, and J. B. Haggin Sucs. to 73,306 hectares, both in the municipio of Janos in Chihuahua.[57] The Babicora Development Company's claim to 101,548 hectares out of the Valenzuela concession in Chihuahua's municipio of Guerrero was approved.[58] But Tyson James's claim to 81,336 hectares in Durango was denied.[59]

Before evaluating the revolutionary state's treatment of the companies, it is first necessary to grasp the state of Mexico's political economy immediately following the revolution. In the first place, the violence had devastated the country's productive capacity. Between 1910 and 1921 output declined in agriculture, mining, and manufacturing. The average annual growth rate of 2.6 percent recorded between 1901 and 1910 was not superseded until after 1936. Mexico's population declined between 1910 and 1921 by 1 million, or about 15 percent, as a result of fighting that both took lives and lowered the birth rate, as well as of increased emigration and an influenza epidemic.[60]

The most lasting effects of the upheaval were not the human suffering and disruption, severe as they were, but the new political and economic institutions established by the generals who called themselves Constitutionalists. The victors in a two-year civil war that pitted them against the agrarian movements of Emiliano Zapata and Pancho Villa, the Constitutionalists stood for much that Porfirio Díaz would have approved of. The Mexican political scientist Arnaldo Córdova noted that they defended the "principle of private property, the plan of capitalist development for Mexico, the institution of a State of law independent of private interests and a legal system of public rights." At the same time, as the words of Mexico's constitution attest, they proclaimed an agrarian reform and benefits for urban workers, social improvements that would be guaranteed by a powerful state.

In short, as Córdova argued, Venustiano Carranza and his allies were the leaders of a *populist* revolution whose principal function was to block the *social* revolution represented by the forces aligned with such leaders as Villa and Zapata—*Se dió el centavo para ganar el peso* (They spent a little to make a lot), as the saying goes. The strategy was implemented by satisfying certain limited popular demands of the masses, whose participation in their own undoing would eventually become institutionalized in a corporativist, paternalistic, and authoritarian state-party arrangement.[61]

Even if Venustiano Carranza's domestic economic policy was little more than an attempt to restore "Porfirian conditions" for the old rich and the new bourgeoisie, as Friedrich Katz has argued,[62] it was one that his successors found congenial. The revolt against Carranza led by Gen. Alvaro Obregón in 1920 owed its success in part to Obregón's popular identification with the more radical demands of workers and peasants.[63] Yet the revolt itself was little more than a barracks coup, mounted chiefly by Sonorans, and it was virtually over in two weeks. Neither the Plan de Agua Prieta, the revolt's manifesto, nor a personal appeal by General Obregón issued the following week, called for the implementation of those provisions of the constitution that would most benefit Mexico's workers and peasants.[64] Obregón, his biographer wrote, was a practical man, "strongly focused on modernization and increased production, not on radical structural change."[65] Of the general's land policy, Eyler N. Simpson concluded that he "can hardly be classified as an ardent agrarian. Indeed, notably in the first years, and to an appreciable extent throughout his term of office, Obregón in dealing with the agrarian problem exhibited a caution which in the face of the high-sounding phrases of revolutionary doctrine and the specific promises of Article 27 amounted to undue timidity if not plain cowardice."[66]

Besides the conservatism of the faction that triumphed, two other factors heavily influenced the new state's treatment of the survey companies: the fragmentation of political power and pressure by the United States to respect U.S.-held properties.

The greatest obstacle in the way of constructing the powerful new state envisioned by the constitution was the dispersion of authority across war-torn Mexico. The army itself, and the factions within the army, as well as mass organizations and regional caciques, were all potential competitors with a national state. Throughout the 1920s numerous caudillos exercised power through regional bosses, military units, and mass organizations, with two negative consequences for the national government, as Rafael Loyola Díaz has shown. The situation not only interfered with and delayed the implementation of national solutions to the grievances that had sparked the revolution. It was also a divisive influence on the fragile coalition that made up the government, giving both the old Porfirian elite and radical forces an opportunity to disrupt the consolidation process.[67]

Reform was also impeded by international law, to which Article 27 was an insolent affront. Implementation of Article 27 would have meant renegade status for Mexico, an underdeveloped nation dominated by foreign capital, and would have invited U.S. military intervention. The U.S. oil industry represented an especially tempting target, and attempts by the Mexican government to regulate its activities had generated a crisis by 1919. Intervention seemed increasingly likely when the U.S. State Department, reacting in part to the kidnapping of the U.S. consul in Mexico, warned Mexico in November 1919 that its failure to make a "radical change in its attitude toward the United States" would "almost inevitably mean war," to quote Secretary of State Robert Lansing. The crisis eased considerably when the oil companies and the Carranza government temporarily accommodated each other.

The subsequent downfall of the "First Chief" was due in part to his anti–U.S. attitude, which Obregón wanted to moderate. Preliminary negotiations began almost immediately with the provisional government installed by General Obregón, the principal source of leverage on the U.S. side being the continued withholding of diplomatic recognition. When Pres. Warren Harding took office in 1921, the U.S. State Department insisted on a treaty that would interpret Article 27 in a "nonconfiscatory, nonretroactive form" as a condition of recognition, and continued to consider the use of military force to protect U.S. interests in Mexico. But Obregón's more tolerant attitude toward foreign capital and his willingness to accommodate the U.S. government helped reduce the pressure for intervention. Several U.S. chambers of commerce accepted invitations from the Mexican government to tour the country.[68] Commissioners from both countries met at the so-called Bucareli Conference of 1923, where the U.S. side expressed its concern about the expropriation of land under Article 27. The United States said it opposed the "provisional occupation of large and sometimes enormous areas of land which . . . are being illegally occupied pending decisions by the National Agrarian Commission" as well as certain "manifest excesses" in the countryside.[69] Treaties were signed providing for the settlement of claims between citizens of the two countries. An informal understanding was reached concerning compensation for land expropriated for ejidos, and Mexico indicated it would respect private subsoil rights acquired before 1917. The United States then recognized the Mexican government.[70]

Taken together, these three factors—the conservative ideology of the Constitutionalist victors, the fragmentation of authority on a national scale, and the intense pressure of an interventionist and aggressive U.S. government (whose troops were roaming across northern Mexico in 1916 in search of Pancho Villa) against full implementation of Article 27—explain both the timing and the timidity of the revolutionary state's efforts to recover survey company land. Table 16 indicates that all revocations but two were issued after 1923, when the Bucareli accords had been signed, the Mexican government had been recognized by the United States, and the threat of intervention by the latter had eased considerably. Because the decrees mandated the cancellation of all titles emanating from the survey company grants, the Mexican government obviously considered it prudent to delay issuing the decrees until relations with the United States had stabilized.

While third-party purchases of survey company land were also canceled by the nullification decrees, the latter invariably gave buyers ample opportunity to seek approval of their titles, as has already been noted. Only U.S. citizens who had invested in the survey companies and who still owned land in the original concessions would have been completely without any recourse. The only company in which U.S. investors were known still to be active, however, was the Sinaloa Land Company—one of the few companies whose holdings were never subject to a revocation decree.

A comparison of Table 16 with Table 9 indicates that the titles to 9 of every 10 hectares given to the survey companies in compensation were nullified by the revolutionary state. Of the 10.5 million hectares titled, titles to 9.7 million were revoked.[71] The eight companies ranking at the top of the compensation list all lost their holdings. Most of those overlooked had been titled fewer than 20,000 hectares; the largest contractor to apparently escape revocation was Joaquín Casasús.[72]

What cannot be determined, however, is what proportion of the land represented by revoked titles was still in the possession of the survey companies. By 1924, more than thirty years after most of the land had originally been titled, much of it had probably been subdivided and sold by the companies, and therefore they would not have been much affected by the government's edicts.

The decrees, by invalidating in one blow all the titles emanating from the controversial survey grants, armed the governing faction and its regional allies with enormous political leverage in the long struggle to consolidate and institutionalize the authority of the new ruling class. All the landowners in the affected areas were forced to appeal to the federal government for a new validation of their right to hold the land they occupied. The revocations, impressive on paper as evidence of revolutionary justice, probably had little impact on the men who profited from the survey provisions of the 1883 colonization law, because it is unlikely that many still held much of the land given them in compensation for their surveys. But the revocations served well enough as responses to pressures from the Left for large-scale confiscations, and above all as instruments of the centralization of power in a fragmented political economy. Lorenzo Meyer's epigram elegantly sums up the official agrarian reformism of post-1917 Mexico: "The Mexican Revolution did not destroy the authoritarian nature of Mexican political life, it modernized it."[73]

7 Summary and Conclusions

This study of Mexico's compañías deslindadoras represents the first attempt to subject long-held hypotheses about the companies to evidence in the archival record. I have also sought to evaluate that evidence within the larger context of a nation ruled by the first politically stable regime in its history, a regime whose development-oriented policies—of which the surveys were among the most prominent outcomes—were to become associated with the decade of violent revolution that followed it. What would a close study of the companies reveal about the political economy of Mexico as the hour of revolution approached?

From the beginning this study has emphasized the extraordinarily influential and (at that time) novel role of the state in shaping Mexico's economic development in the three to four decades preceding the revolution. Its commitment to following the path taken by the more developed capitalist countries was demonstrated throughout the period, as the governments of Manuel González and Porfirio Díaz sought to improve communications and transportation, to open the public lands to development, to institute the legal norms and procedures that would encourage capitalists (especially foreigners) to invest in Mexico, and to keep the peace.

In the realm of the disposition of the public lands, the state's role was clearly defined from the outset. The imperatives of raising revenue and stimulating economic growth called for swift action to identify public land and sell it to individuals or companies, thus enriching the public treasury and fostering economic growth. In addition, individuals who held public land illegally would be forced to take steps to perfect their titles to it. Unleashing private surveying companies armed with a federal authorization was considered the quickest and most economical means to these ends. It was also politically shrewd to hire private contractors to do a job that was sure to provoke opposition from landholders rich and poor. The companies, in Díaz's apt phrase, were both instruments and shields—cat's paws that could be disciplined or tossed aside whenever popular ire mounted in a given locality.

The conclusions of the preceding chapters have been presented against the backdrop of a historiographical tradition that made the survey companies the instruments of large landowners who sought to dispossess peasants and communities to "monopolize" even more land, which was usually not productively utilized but simply held for "speculative" purposes. This caricature has long been a tempting target of historical research, but the lack of ready access to the Archivo de Terrenos Nacionales and the frequency with which the legend is still repeated by present-day historians allowed it to linger beyond its natural lifetime, which one could define as the period prior to the 1960s when the agrarian tradition of Mexican revolutionary historiography began to be seriously challenged.

The legend of the survey companies has acquired such authority that it is rare to find works that even bother to cite evidence of company wrongdoing.[1] I have not sought to create a contrary legend, but rather to submit the question to the obvious source—the repository of data generated by the surveys themselves—and to attempt a balanced interpretation of the evidence. Therefore some obvious, and not so obvious, caveats to the general interpretation presented here need to be stated, and in some cases restated.

First, like all sources, the records of the Secretaría de Fomento are incomplete and frequently misleading; the absence of evidence of a protest, an illegal land seizure, or a government decision impinging on landholders' rights is not proof that such events did not occur, as I noted in the Preface. Local judicial authorities were probably no less tempted to submit to a *mordida* (bribe) offered by a real estate operator than judges in, say, Chicago in the 1980s. Surveyors surely seized land that was being worked, with legal title or without, by rural producers; the high proportion of indeterminate cases among those in which some opposition was expressed has already been noted. Others who lost land may never have submitted a complaint, out of ignorance or poverty. Documents alleging misconduct by the surveyors may have been lost or destroyed, perhaps intentionally.

Second, the existence of formidable natural hazards in a relatively impoverished and technologically backward country undoubtedly undermined the accuracy of even the most scrupulous survey. As I noted in chapter 2, José Covarrubias, the head of a Fomento section in charge of the terrenos baldíos and a persistent critic of the companies' performance, estimated it would take more than thirty years for one team to produce a truly accurate survey of Chiapas, pointing out that the dense jungle and the eight to nine months of rainfall would make even that a difficult feat. One government surveyor in Tabasco reported that even during the dry season his men could not walk more than fifty meters from the banks of the rivers without sinking up to their waists in mud. Hostile Indians threatened surveyors in the northern states.

Moreover, the legislation authorizing the government to contract with survey companies was repealed in 1902 in part because abuses did occur. The inaccuracy of many of the surveys and the erroneous identification of privately held land as baldío were among the factors that led to the repeal, as we saw in chapters 2 and 6. In 1909 Díaz went further, suspending the sale of all surveyed public land until it was remapped by official government survey teams, a measure required by the uncertainty of the government's knowledge of the public domain, according to Fomento chief Olegario Molina.

Finally, it bears repeating that the companies hired to survey the public land were not the only entrepreneurs seeking to acquire public land. Thousands of others were potential sources of disruption in the countryside. More than seven thousand titles covering about 17 million hectares in public land were issued between 1878 and 1908 to companies and individuals who filed claims under the 1863 public land law, and who were usually required to submit surveys of their claims. Another 5.2 million hectares were sold outright by the government after they were surveyed. Other cases of usurpation occurred that had nothing whatever to do with the disposition of the public lands. In any case, that the confiscation of smallholders' land was a prominent grievance before and during the Mexican Revolution has not been disputed here. The methods by which land could be stolen during the Porfiriato were many and varied, and the deeply rooted agrarian sources of the great conflict cannot be ignored.[2]

The regimes of González and Díaz—so closely identified that they are usually lumped together as the Porfiriato—directed the affairs of a relatively autonomous state apparatus. The state ruled on behalf of the emerging entrepreneurial elite (as represented in this case by the survey contractors), but far from being their instrument, the state itself manipulated them for its own purposes. While they were members of this elite, the survey companies were often locked in bitterly contentious conflict with the state over such issues as land usurpation, the accuracy of the surveys, the division of surveyed land between the company and the state, and adherence to various legal and technical requirements. These conflicts were often settled when the state acted to discipline the contractor by nullifying land titles, suspending authorizations to survey, or simply by instructing the company to take (or not to take) certain actions.

The Díaz state recognized that a diversity of interests was at stake in the surveys and understood that interests other than those of the contractors sometimes had a higher claim. Peace in the countryside was considered essential to attract foreign investors, and that made political stability the keystone of the Porfirian state's development strategy.[3] The threat to stability posed by attacks on both nontitled occupants and legitimate landowners at times induced the Secretaría de Fomento and President Díaz personally to act to protect those rights. Preventive measures were taken to defend landholders, whose grievances against the companies often resulted in action by Fomento on behalf of the complainants. The state, as Porfirio Díaz explained in a letter to the governor of Morelos (see chapter 1), found it easier to exercise this discretionary authority precisely because private companies, as opposed to official state agencies, were hired to do the surveys.

The companies recognized that respecting the claims of property holders was often cheaper than engaging in litigation or administrative processes that could delay the issue of a title for years, not to mention the risk of physical attacks on their engineers by offended landholders. As noted earlier, the archival record indicated that the companies generally respected the claims of property holders. It is likely, however, that some landholders were despoiled of their property. In such cases, the victims were probably so powerless (i.e., lacking in organization or other resources) that they were unable to threaten the companies sufficiently to deter them from taking their land illegally. Without a clear title to the land being claimed, a community or property holder was vulnerable to having the land identified as baldío and losing it, but sufficient legal claims could still be made (on the basis of prescriptive rights, for example) to keep a survey company at bay, provided the resources for making such claims were available.

But just how relevant were laws and the law? Frequent reference to the law has been unavoidable in a study whose principal sources constantly cite land titles, land laws, court petitions, legal claims, judicial verdicts, and administrative decisions based on laws and regulations.

Weighing the legality of any particular event opens one to attack by those who question the relevance of law to the history of social changes such as those being investigated here. Their argument is more or less as follows. Because the laws and constitutions of nineteenth-century Mexico were by definition written and then enforced (or not enforced) by the ruling class, group, or formation, they were little more than instruments of class rule. Whether a particular survey company followed the rules matters little, because the rules were written by those who stood to benefit most by the policy in question. All that counts is the outcome: Who lost and who won? To a small farmer or a peasant community, the loss of land or water had the same effect whether conducted legally or illegally, and the distinction between the two is irrelevant.

This position assumes that nineteenth-century peasants, either through their own ignorance or as a result of the machinations of the ruling class, were essentially excluded from a property system that normally required evidence of ownership in the form of a land title issued and validated by some political authority. There are several reasons, however, to challenge this argument.

First, peasants were not unacquainted with land titles, which had been introduced by the Spanish imperial authorities and perhaps even used before the Spanish conquest, as William B. Taylor found in Oaxaca:

> Villages and individual [Indian] nobles frequently petitioned the viceroy for formal titles and forged titles when their rights to untitled lands were challenged in the seventeenth and eighteenth centuries; the apparently close link between land tenure and pre-Hispanic writing systems in Oaxaca and central Mexico meant that at least some Indian societies had had experience with written entitlement before the conquest.[4]

A similar pattern existed during the colonial era in the Valley of Mexico, according to Charles Gibson:

> Throughout the colonial period the native community prized and guarded its land in full awareness of the dangers of alienation. In the first post-conquest years, as in the late Aztec period, land disputes between one town and another were a standard feature of Indian life. But in the developed colony, though controversy within Indian society continued, the more characteristic disputes occurred between Indians and Spaniards. The communities' ideology of land protection was expressed most forcibly in their 'títulos,' documents of uncertain origin, of indisputable native composition, of considerable persuasive power, but of limited legal validity.[5]

Both Taylor and Gibson found that peasants were also accustomed to defending their rights within the legal system set up by the conquest.[6] There is no reason to think that respect for land titles or the willingness to litigate declined between independence and the Díaz era. Given the chronic political instability of the country in that period, reliance on legal forms may have become even more commonplace as a defensive measure.

The *deslindes* (surveys) files of the ATN consulted for this study indicated that any threatened landholder or community's first recourse was to some kind of written evidence of ownership or at least possession. Only when such a document could not be produced or at least cited (petitions frequently claimed that the title had been destroyed or lost) was an argument for prescriptive rights mounted. The historical literature fairly overflows with references to communities of poor farmers petitioning the state to recognize their legal rights to the land they claimed on the basis of ancient titles. The very sense of legal right that permeates these documents shows just how pervasive was respect for the law and legality in matters related to land. The Mexican peasantry's adroitness in manipulating the legal system has scarcely diminished over the years, as the struggle for land in the Huasteca region during the 1970s illustrated.[7]

If peasants were not only familiar with the meaning of legal entitlement and with defending their rights in the courts but also had lengthy experience with titles and their defense long before independence, the issue of legality is hardly irrelevant, any more than legality would be irrelevant in a dispute between, say, two European-educated capitalists over urban property.

Indeed, even in the agrarian conflicts of eighteenth-century England, in what one might think of as the very antipode of Mexican "culture," E. P. Thompson has made a very similar argument against Marxist structuralists who would define law as little more than another ruling class facade. The rulers and the ruled struggled not over law but over the laws themselves, in the English case over alternative definitions of property rights. Law was less often a superstructural excrescence than a matter of how one defined "actual agrarian *practice,* as it had been pursued 'time out of mind.'" This observation by Thompson could as easily have been uttered about nineteenth-century Mexico. People, he concludes, "are not as stupid as some structuralist philosophers suppose them to be. They will not be mystified by the first man who puts on a wig." Moreover, if the law is so abused by the lawmakers that it is widely perceived as no more than a ruling class instrument, it is worse than useless. Its evident injustice will subvert its very purpose; "it will mask nothing, legitimize nothing, contribute nothing to any class's hegemony."[8]

For hundreds of years rural Mexicans of all classes understood the significance of a title to land issued by the state. But by the late-nineteenth century, the "verdadero caos" (real chaos) of a rural property system where boundaries lacked the sharp definition required of the age of capital forced the state to undertake scientific surveys of the public lands.[9] Some of these lands were undoubtedly being exploited illegally (i.e., without title). To dislodge squatters the Díaz and González governments resorted to ancient custom by insisting either on documentary proof of a right to occupancy, or at least evidence of customary usage. The survey companies became the means by which squatters could be pressured to prove their right to the land they occupied. If they could not produce a title, other alternatives were available: showing evidence of customary usage, filing a claim under the law of 1863, or arranging a composición with the surveyor. Numerous landholders, confronted by the surveyors, took advantage of precisely these alternatives. Mexican governments, as was shown in chapter 1, had been grappling with this problem for decades, reasoning correctly that unless threatened with eviction, squatters would have little motivation to comply with the law.

If the occupants of public land lost it during the Porfiriato because they never had a legal right to the land in the first place, then those who finally acquired such land—the survey companies, for example, and those to whom the companies transferred it—could scarcely be called usurpers. In the eyes of the state, it was the victim of usurpation by the occupants. This may be one reason why resistance to the companies by landholders was disproportionately slight. Landholders in such a situation had little alternative but to submit to the law. In any argument with the survey companies over land, the single most important piece of evidence was a written title. When proper titles were produced, they were almost invariably respected. When they were not produced, the survey companies' claim that the land was baldío was difficult to dismiss; the risks of litigation in such cases must have risen astronomically, because the probable returns to the petitioner were extremely low. Submission was often the wisest course.

Historians no longer attribute the revolution mainly to a single source, such as agrarian grievances or the exclusive character of the Díaz regime. The causes were extremely diverse, and the contribution of the regime's public land policies to the discontent that helped drive the revolution is difficult to assess. Chapter 6 concludes, on the basis of archival research and on the secondary literature that touches on the subject, that the Díaz government's land policies—including the claims process and the division of ejido, as well as the use of the survey companies—scarcely contributed to discontent in the countryside. Expropriations carried out by individual landowners acting well outside the sphere of public land policy appear to have been the principal sources of grievances over land.

Apart from the surveys' impact on participation in the revolution, they eventually became a prominent part of the post-1920 politics of the revolution. The victorious Constitutionalists revoked most of the companies' concessions and land grants. The revocations, however, were probably less effective in recovering land given to the companies than as weapons in the conservative, procapitalist regime's struggle to consolidate and institutionalize its authority across the country, as chapter 6 concludes. It is unlikely that much land remained in the hands of the survey companies by the 1920s. Chapter 5 pointed to a pattern of fairly rapid subdivision and transfer of company land grants following entitlement, and the postrevolutionary government's revocation decrees always respected land acquired from the companies by third parties, though even they were required to justify their claim. The decrees served the new revolutionary state headed by Alvaro Obregón and Plutarco Elías Calles by providing it with a politically unassailable instrument for enlarging its power, by putting pressure on foreign and domestic landowners and local authorities to justify their landholdings or (in the case of local politicians) those of their clients. Implementing a genuinely revolutionary program of land redistribution was not among the objectives of a regime that sought the consolidation, as Adolfo Gilly noted, "of a new national bourgeoisie, fundamentally using the level of the State to affirm its domination and to assist it in the accumulation of capital."[10] The Mexican revolutionary state, led by a rising new bourgeoisie that understood the necessity of integrating peasants and workers into the state itself (by means of Articles 27 and 123 of the constitution of 1917), was more a refinement of Porfirismo than a rejection of it.

Finally, how does one explain that, during a period of mounting rural discontent, the survey companies generally did not become the blunt instruments of a land-expropriating elite?

The state's capacity to intervene against land-expropriating hacendados was considerably narrower than its capacity to intervene against the survey companies, which after all were agents of the state. In conflicts between large landowners and the rural poor, Díaz may have preferred to provide more protection to the latter than he actually succeeded in doing, as John Tutino has speculated. But it was politically unfeasible to side with the poor in such conflicts, since Díaz's survival in the presidency required him to allow rural elites to enrich themselves at the expense of the peasantry, if necessary.[11] It was, however, politically acceptable—even necessary, as noted earlier—to exercise comparatively rigid control over the activities of the survey contractors, agents of a state that was paying them generously for their services.

This behavior can be attributed to the dilemma that the state confronted when it adopted the survey policy. On one hand, the survey of the public lands was a step toward building a modern capitalist economy and was needed to attract new investment in Mexico's natural resources. On the other hand, hardly any rural producer of whatever class who was peacefully exploiting land to which he was not legally entitled, or whose titles were vague or of questionable legitimacy, had any reason to welcome a survey. A survey was an act of aggression no matter how fairly or accurately it was conducted. The Porfirian state therefore adopted a strategy that gave it the flexibility it needed to absorb both demands simultaneously. The companies, as agents of the state, would survey the land, but they would also be held accountable for the inevitable opposition that developed in the countryside.

The survey strategy provided a means to balance the diversity of interests at stake in the transition to capitalism, a diversity that required a certain measure of state autonomy from both the modernizing sector and the landed elite.[12] After most of the public lands were surveyed, Díaz abolished the contract system and replaced it with surveys run directly by the government, few of which were ever undertaken, as we have seen. The companies had served their purpose.

Historians of the period have not sufficiently appreciated the complex character of the Porfirian state, particularly its capacity for autonomy. President Díaz may have been reluctant to restrain the arbitrary and expansionist behavior of the hacendados, though the evidence on this point is not yet conclusive. But in the case of the survey companies the regime nevertheless managed to establish a crude system of regulation and oversight as it sought to strike a balance (necessarily incomplete and inconsistent) among conflicting class interests while serving goals considered beneficial to Mexico's long-term economic development, such as accurate surveys of rural properties and the prevention of class violence in the countryside. The increasingly sclerotic and autocratic administration of Porfirio Díaz ultimately failed to deliver on both counts. But the archival record not only undermines the traditional linking of the two failures. It also points to a complex and dynamic interplay of state and private interests that has much less to do with the regime's demise than with its early resilience and remarkable longevity.

Appendix A

Twenty Survey Enterprises

Profiles of twenty enterprises that held survey concessions in the six-state study area follow. The companies' profiles, presented alphabetically, include all relevant biographical and financial data that could be found in three Mexico City archival sources and in contemporary as well as historical published accounts.

Antonio Asúnsolo

The survey concession for Chihuahua and Durango awarded to Antonio Asúnsolo in July 1884 was by 1890 owned by a company in which Asúnsolo himself only held 20 percent of the shares. Tomás Macmanus and José Valenzuela also held 20 percent each, while Telésforo García owned 40 percent. The Banco de Londres y México bought Macmanus's share in 1896, and in 1904 García acquired Valenzuela's. García sold his shares to Leopolo Iwonsky, who sold them to the bank and Asúnsolo's estate in 1906, and in 1909 the bank and Asúnsolo's estate were still the sole owners.[1] Asúnsolo also invested in several other companies, as will be seen later.

José María Becerra

The Becerra family of Chihuahua, political allies of the anti-Díaz strongman Luis Terrazas, controlled the municipalidad of Urique in the district of Andrés del Río. Brothers José and Buenaventura dominated the area's mining industry as well as its politics, enriching themselves in the process. José died in 1889 while in London on business. The brothers' association with a prominent opponent of Díaz appeared not to damage their standing with the president, however. Not only was José awarded a contract to survey in Sinaloa in 1886, but his brother represented Chihuahua as a *suplente* (substitute) in the Senate in the 1890s.[2]

Among the investors in Becerra's concession was Mariano Martínez de Castro, who told Fomento that he withdrew from the firm after he was elected governor of Sinaloa in 1888, "as a prudent measure" to avoid "any fear of inappropriate influence on my part." A Sinaloan landowner who was critical of the company's behavior (see chapter 4) claimed that other investors included Luis Martínez de Castro, the governor's brother and the man who would eventually monopolize the public land survey business in Sinaloa after 1900, and Manuel Monzón, a local judge.[3]

Just two years after the Becerras received their contract, they tried to transfer it to a U.S. citizen, Lucius Clark, a partner in the government's railroad concession to the Ferrocarril Mexicano del Pacífico. All such transfers required Fomento's approval, and this one apparently was rejected. In July 1888 it was sold to Luis Gayou, a Mexican citizen residing in Guaymas, with President Díaz's approval.[4] Gayou conducted several composiciones but failed to fulfill certain unspecified obligations to the Becerras, who sued and got the contract back in November 1889. They immediately requested, and received, permission to transfer it to Joaquín D. Casasús.

Under the agreement with Casasús, all the titles received by the Becerras in the district of Fuerte would remain theirs. But land owed on surveys done in the rest of the state would be Casasús's, who was to refund to the Becerras all the costs already incurred in those surveys. As a guarantee that he would pay them, Casasús deposited $3,000 in the Banco de Londres y México. The Becerras also reserved the right to survey in the directoría of Choix, Fuerte district, in those properties designated by Doña Natalia Valenzuela de Saracho (except for a finca owned by Carlos Flurrnoy). The Becerras would also have the right to any *huecos* (unaccounted-for land) or demasías found on their own land and that held by Cosme Almada, as well as the right to survey any land that Governor Martínez de Castro might designate on three properties in the district of Culiacán. Casasús paid 6,370 "pesos de oro americano" for the contract, which he purchased on behalf of the U.S. citizen William F. Robertson of Eagle Pass, Texas. Three months after he bought the contract, Casasús sold it to Robertson for the same price.[5]

When Fomento inquired of Robertson's nationality, Casasús explained that he was a wealthy U.S. citizen who had built nearly all of the Ferrocarril Internacional de Mexicano between Piedras Negras and Torreón. Yet Díaz did not approve the transfer to Robertson until January 1891, a year after the American bought the contract. Robertson hired Pablo and Manuel Martínez del Río to represent him and won a two-year extension of the contract from the president. Díaz, meanwhile, turned down Buenaventura Becerra's request to continue his right to survey in Fuerte district.[6]

The contract changed hands again in 1894, when William H. Mac-Wood acquired it, but there is no evidence this transfer was ever approved, nor is there evidence that either man received any land in compensation for surveys they conducted. Robertson may have bought the concession for the rights to land due Becerra for various surveys and composiciones that were pending.[7] By 1899, however, Manuel Martínez del Río was calling himself the "assignee" of the Becerra contract and appeared to have acquired all the rights to it by then.

As noted earlier, Becerra had transferred his rights to survey in the directoría of Choix to Doña Natalia Valenzuela de Saracho, the sister of José Valenzuela and wife of Doroteo de Saracho, the managing partner of the Sociedad Minera Doroteo de Saracho y Compañía. The latter was seeking land to build an ore-processing plant in Choix in the early 1890s.[8]

Joaquín D. Casasús

Of all the men associated with the survey companies, Joaquín D. Casasús was the most prominent member of the "científicos," the inner elite at the top of the Díaz government. A Tabasco-born lawyer, businessman, economist, and translator of literary classics, Casasús was early on a fervent political supporter of Porfirio Díaz. He became one of the "trinity of lawyers," in Daniel Cosío Villegas's felicitous phrase, who along with Pablo Macedo and Rosendo Pineda made up the *capa superior* (upper level) of the científicos. Casasús advised the government, represented it abroad, and joined it as a member of the Cámara de Diputados (1886–1903) and a senator (1907–11). While accumulating a sizable fortune as a lawyer, Casasús became with Macedo one of the "caudillos of the economy in the apogee of Porfirismo."[9] He backed Díaz to the very end, dying in exile in New York in 1916.

One contract was awarded to Casasús in 1887 to survey four districts in Durango. He also acquired José María Becerra's contract to survey in Sinaloa in 1889, which he then turned over to the U.S. citizen William T. Robertson three months later (see Becerra profile). Thus Casasús only conducted survey operations under the Durango contract and in compensation was titled 414,611 hectares of land in Durango. This sum was reduced by half when two of the titles were nullified during the Díaz era, for reasons that are not recorded.

Casasús's name was only on one of the seven titles issued in compensation for his survey, however, and it turned out to be one of the two that were later nullified. The other titles were issued directly to third parties. There is no evidence that the title holders were owners of the company; Casasús merely requested that the titles be issued to "different individuals with whom private arrangements have been made" to quote one of the titles.[10]

Casasús represented Enrique Creel in 1910 in the latter's suit against Jesús Valenzuela's estate to recover a loan (see the profile on Valenzuela's company).[11] He also represented the Sinaloa Land Company in 1910 and Policarpo Valenzuela in the latter's conflict over his surveys with the government of President Francisco I. Madero.

Compañía Descubridora de Terrenos Baldíos

After Fomento issued four separate authorizations to survey in Durango to Antonio Asúnsolo's company (July 1884), to Meza Hermanos y Compañía (March 1882), to a partnership formed by Telésforo García and Jesús Meraz (July 1883), and finally to Telésforo García alone (February 1883), the contractors merged their concessions in December 1884 under the management of Asúnsolo. The step was taken, Asúnsolo's agent explained, because the company "ran up against the difficulties which the hostile interests and aspirations of the other survey companies opposed. In order to smooth them out and avoid trouble and obstacles," Asúnsolo "managed to combine all the authorizations that he could." The various concessions together authorized the survey of virtually the entire state.[12]

The ownership of three of the four concessions overlapped. Not only did Telésforo García hold two in his own name, but he was also an investor in Asúnsolo's company. (See the profile on Asúnsolo.) Moreover, both Asúnsolo and García were investors in the concession held by García del Campo and Ramón Guerrero (see profile).

The concession held by Meza Hermanos (via Gen. Remedios Meza) attracted the single greatest participation by the Mexican political elite in any survey company in the study area, yet it appeared to be one of the least successful. Although originally awarded to Meza Hermanos, it was quickly acquired by the Compañía Descubridora de Terrenos Baldíos en el Estado de Durango during the González administration. The company's stockholders included none other than Carlos Pacheco, the secretary of Fomento when the company was organized and received its contract, and Porfirio Díaz himself, who was governor of Oaxaca when he purchased the shares. Pacheco and Díaz each held only three shares out of fifty, the six being recorded in the name of Jorge Hammeken Mexía,[13] the editor of the daily *La Libertad,* whose fate it was to be ruthlessly abased by the authoritative historian of the Porfiriato, Daniel Cosío Villegas. Don Jorge "was an individual bereft of any political significance and, apart from being a poor writer, he had no other gifts than to be a compatriot and a friend, and now [during the González administration] to manage the private businesses of Porfirio in the Capital."[14]

Besides Pacheco, Díaz, and Hammeken (who held three shares of his own), other investors included Manuel Romero Rubio, Díaz's newly acquired father-in-law and his secretary of *Gobernación* from 1884 until his death in 1895; Ramón Fernández, President González's principal adviser, and an investor in one of Rafael García Martínez's survey companies; Alfonso Lancaster Jones, a deputy; and Gen. José Montesinos.[15] None of the above held more than three shares (out of a total of fifty), which at ten pesos each hardly appears to have been much of a risk. In April 1883, a year after the company received its contract, the board of directors consisted of Ramón Fernández, president; Romero Rubio, vice president; Joaquín María Escoto, secretary; Gen. José Ceballos (hero of the war against the French, who would be Díaz's governor of the Distrito Federal from 1884 until 1893), the alternate secretary; Gen. Fernando Poucel, treasurer; Gen. Enrique Ampudia, alternate treasurer; and Gen. Sostenes Rocha, Lancaster Jones, and Hammeken Mexía.[16]

The Meza concession did not prosper, however. In June 1884 Romero Rubio, now the company's president, requested and received a two-year extension of the contract. Finishing the job in three years would be "impossible," he told Fomento, owing to the size of the area to be surveyed as well as the "endless difficulties of all kinds that arise from the resistance mounted by many of the proprietors that border with them [the land surveyed] and occupy and retain them. The reality and importance of such difficulties is well known to the Government because they are common to this type of business."[17] Thus even a company whose ownership was part of the nation's ruling clique found its work practically paralyzed by the opposition of property holders, a circumstance that speaks not only to the occupants' readiness to resist a survey but their ability to do so as well.

As a result of the company's lack of progress in conducting the survey, Romero Rubio addressed an extraordinary meeting of shareholders in July 1884, two years and three months after the firm was organized, by which time it was also facing a serious capital shortage. For it to continue in the "present undefined situation, without subscribing the necessary capital, nor carrying out any operations of any kind that would lead to putting this enterprise into business, is not only prejudicial" to the persons that formed it and to the nation, Romero Rubio said, but it also hurt the "good name and honor of each and every one of the stockholders, inasmuch as its conduct before the eyes of the government and the public may be subject to criticism at least regarding the neglect of the solemn commitments for which they contracted with the country upon accepting the concession that was granted to them."

He proposed two options: either raise the capital necessary or sell the concession. As is generally known, he added, "the owners of other government concessions, equal or analogous to that [concession] which this company has, have signed contracts with capitalists of known competence, transferring to them their obligations and rights, but reserving one-third of the profits that result from the surveys that are carried out, free of any cost or encumbrance."

The shareholders unanimously opted for the second option, that is, selling the concession, and Romero Rubio was authorized to either contract with a company or "unas capitalistas," or to transfer the concession, if necessary.[18]

Later that day the two Chihuahua survey contractors, Antonio Asúnsolo and the firm of Macmanus e Hijos, signed an agreement under which they would take over the management of the concession for two-thirds of all the land the company was to receive under it, the balance to be kept by the company. Asúnsolo and Macmanus would pay all the expenses of survey, including "payments of honorariums either to lawyers or to other persons that may be needed to obtain the results sought after." The company would have to deal with Fomento and obtain the titles. Neither Asúnsolo nor Macmanus would be held responsible if the concession expired because of "incursions of barbarians or other insuperable [causes]." Asúnsolo and Macmanus paid the company $1,500. The agreement gave the two no equity in the company, but four days later Gen. Ceballos sold his single share in the company to Asúnsolo for $1,000—a tidy profit of 9,900 percent. Ceballos could buy it back within a year for the same price plus interest. Asúnsolo also acquired Telésforo García's stake (unspecified) in the concession he held with Meraz.[19]

García and Meraz then joined in, uniting all the concessions under a new company called the Compañía de Concesiones Unidas de Deslinde de Terrenos Baldíos en el Estado de Durango. Later, Jesús Meraz sold his share in García-Meraz to Asúnsolo. Asúnsolo obtained still another survey concession in July 1884 for the new company. It finally surveyed 1,043,099 hectares in the partido of Santiago Papasquiaro, for which title to 347,699 hectares was issued by Fomento in Asúnsolo's name in September 1887.[20]

The shares that belonged to Lancaster Jones, Generals Rocha and Poucel, Escoto, and Carlos Quaglia were bought by Carlos Guerrero between 1896 and 1907. In 1900 Romero Rubio's widow sold his three shares to Luis García Teruel. (Both Quaglia and García Teruel had obtained rights to the expired Jecker survey concession, described below.) In a single transaction in March 1907, Guerrero then ceded his twenty shares to the

Banco de Londres y México and to Asúnsolo's estate, and García Teruel did likewise with his three. Their price: $1,500 a share. Finally, the bank bought the 7½ remaining shares held by the family of General Meza in 1906.[21]

During the next four months Guerrero acquired the late general Montesinos's two shares from his estate, and 2⅒ of Hammeken y Mexía's remaining original three shares. He sold them all to the bank and Asúnsolo's estate in July 1907 for a total of $2,640, or only $643.90 a share— a 57 percent drop in value in just five months.[22]

Telésforo García

The name of Telésforo García, a Spaniard who founded the capital daily newspaper *La Libertad,* appeared with more frequency in Fomento's survey company archives than any other contractor or company agent, showing up in every state studied except for Tabasco. García had opposed Porfirio Díaz's revolt of 1876, but he was soon converted. After he and the editors of his newspaper accepted a subsidy from "amigos del gobierno," García abandoned the management of the paper, "tempted by the proximity of power," and maneuvered himself into a strategic position at the public trough that was so generously replenished during the administration of Manuel González. García could have stayed in his ivory tower, "but with the birth of the great economic expansion of the González Era, within García bloomed the Spanish businessman that had been asleep, so that he came to achieve distinction as one of the great beneficiaries of the epoch. So involved did he seem to be in the González administration that he could not convert quickly enough to situate himself adequately in the new administration of Díaz." Shortly after Díaz's second inauguration, García's name was among those prominently associated with the corruption of the González administration by those who most wanted to "demonstrate their porfirismo." A brief self-exile in his native Spain was García's prudent response.[23]

All five of the survey contracts issued in García's name were obtained within the first seven months of 1883 during González's administration. He sold three of the five to Luis Huller, without ever having received land in compensation for surveys conducted under any of the five. In only one of the five contracts was evidence of other investors found. García's partners in a survey contract for part of Sinaloa (later sold to Huller) included J. Enriquez, Enrique Mackintosh, and J. Sierra. The second was a partner in Jesús Valenzuela's company, and the third was undoubtedly Justo Sierra, one of the age's outstanding intellectuals and an editor at García's newspaper.[24]

García also invested in other survey companies. In 1890 he held 40 percent of Antonio Asúnsolo's company, shares that he still possessed in 1905 when the government was forced to auction some of that company's land because it was unable to pay the penalties for failing to fulfill its colonization obligations (see chapter 3). García held the largest single stake in the concession acquired by Jesús Valenzuela to survey in Chihuahua in 1883 (see Valenzuela profile), by which he received title to 189,418 hectares in Cantón Jiménez, issued by President Díaz in 1885. By buying an IOU held by Manuel Aguayo against Lauro Carrillo for three thousand pesos advanced to cover the expenses of Carrillo's survey in Sonora in 1887, García got 78,332 hectares there in 1899.[25] In addition, he purchased Díaz González's contract to survey in Chihuahua in 1883, which he later transferred to Jesús Valenzuela. He also bought Mariano Gallego's contract to survey in Sinaloa.[26]

García represented Macmanus e Hijos at Fomento in their survey of Sonora. His brother Mariano acquired Eduardo Valdez's contract to survey in Sinaloa, which he later transferred to Luis Huller. Mariano also paid $256,991 for 395,372 hectares in Cantón Degollado, Chihuahua, under a colonization contract in 1885, on land surveyed by Valenzuela.[27]

García, even though his companies did little surveying, was assured of cooperation at the highest levels. More than a year after he was awarded contracts to survey in Baja California, Sinaloa, and Sonora, President González wrote to Gen. José G. Carbó, "a veritable military proconsul" in Sonora and Sinaloa who was chief of the northern military zone. García would start his survey soon, the president said, "and since for that purpose he needs your valuable and effective assistance, I beg you to provide it to him unstintingly, so that he may not encounter any problems or undue delays in carrying out his activities." Carbó immediately assured García of his "trivial cooperation in assisting him" in the survey, and advised the alcalde of Mazatlán to cooperate.[28]

Rafael García Martínez

García Martínez received two contracts to survey in Durango while he was *jefe de órdenes* on the president's general staff, a position he held from 1880 to 1894, when he was promoted to colonel and named jefe político of the southern district of Baja California.[29]

In November 1883, four months after García Martínez was authorized to start surveying in Durango, he, Carlos Ruiz, and Celestino de la Quintana organized a company called La Esperanza. The company's charter specified that of the terrenos baldíos the company would receive in compensation under the survey contract, 10 percent would be set aside for the people "who have lent their services and assistance to obtain the concession," and 90 percent would be divided into ten shares, distributed in the following proportion: five for García Martínez (the managing partner and administrator), three for Ruiz, and two for Quintana.[30]

By 1890 the 10 percent had been distributed before a Durango notary. Ruiz, García Martínez, and the estate of Quintana (who had not lived to enjoy the fruits of the contract) then agreed to dissolve La Esperanza and to distribute their own shares in the agreed-upon manner. In 1890 the company held 278,006 hectares in Mezquital and 5,119 hectares in San Dimas, in land given in compensation for the surveys, which had been titled either to García or the company. (For the division of the Mezquital and San Dimas lands, see chapter 5.) The three parties also agreed that since García Martínez was accepting responsibility for the company's remaining liabilities, he would get exclusive ownership of any baldío the company was still due in Durango, regardless of the area or the value of the land.[31]

La Esperanza was not the only company García Martínez organized to exploit his concessions. In April 1888 he and Hubert H. Warner formed the Compañía Mexicana Limitada Deslindadora de Terrenos Baldíos, for which García obtained "diversas concesiones" from the federal government. In return, Warner provided the capital. But "some difficulties" occurred "among the partners," leading Warner to file a court suit demanding the nullification of the company's charter and the liquidation of its operations, restoring to each partner his original contribution. In October 1889 the judge approved an out-of-court settlement under which Warner would be reimbursed, and would also be entitled to buy from García Martínez land owed Warner from the survey of Mezquital by La Esperanza, as well as land due him as a result of the new company's survey of property called Pinos Altos in the partido of Durango. The price of the land would be the balance in Warner's favor after the liquidation of the company.[32]

Warner had contributed $27,856 between April 11, 1888, and August 10, 1889, which after deductions shrank to $23,866 due Warner, according to the liquidator's report. The company was dissolved in March 1890, and all of its remaining assets and liabilities remained García's responsibility. Instead of paying Warner his cash due, García turned over to him the 88,062 hectares in Mezquital that he had just been given upon the dissolution of La Esperanza. García also gave Warner 7,690 hectares in Pinos Altos. The two lots extinguished his debt to Warner. García had the right to buy back all the land within four years, at the same price plus 6 percent a year.[33]

García Martínez organized yet another company, this one called the

Sociedad Rafael García Martínez y Socios, to finish the surveys in April 1892. This time the partners, besides García Martínez, were Andrés Horcasitas, a Chihuahua lawyer and judge; Ramón Fernández, the adviser to President González who had also invested in Remedios Meza's Durango concession (see Compañía Descubridora de Terrenos Baldíos); and Epitacio Ríos. The latter would finish the survey at his own expense, while Horcasitas, a circuit court judge in Mexico City, would grease the wheels in the capital, obliging himself to

> efficiently arrange the rapid and profitable management of the deals pending at the Secretaría de Fomento, and most especially those awaiting the attention of the appropriate minister in order to smooth out any difficulty that might present itself in what concerns the government and also to provide his assistance as a professor of law in the judicial matters that might arise during the course of doing business.[34]

In exchange, the "proceeds that may arise from the business" would be divided thus: 59⅙ percent for Rios, 28⅓ percent for García Martínez, 8½ percent for Horcasitas, and 4 percent for Fernández.[35]

Ignacio Gómez del Campo and Ramón Guerrero

These two men held two of Chihuahua's four seats in the Cámara de Diputados from 1880 to 1886. In 1882 they both received a contract to survey three cantónes of that state, as well as two municipalidades in the cantón of Abasolo. A year later Gómez del Campo alone was authorized to survey in Sonora. His brother Patricio held a separate contract to survey in Chihuahua, which was voided in 1886 because no work was ever done under it.[36]

Investors in the concession held by Gómez del Campo and Guerrero included Francisco Macmanus e Hijos, a Chihuahua merchant family that also held a Sonoran survey concession, and whose business went backrupt in 1893; Antonio Asúnsolo, who also had a survey contract of his own; and Enrique C. Creel, the Chihuahuan banker. When Patricio requested his authorization to survey, he also identified Asúnsolo and Tomás Macmanus as partners in his own ill-fated venture.[37] Creel was a close political confidant of President Díaz and kept the president well informed of events in Chihuahua; he was heavily involved in the choice of Lauro Carrillo to succeed Gov. Carlos Pacheco, when the latter was named to head Fomento. Neither Guerrero nor Gómez del Campo, however, apparently enjoyed Díaz's full support when the president retook office following the González interregnum. Neither man served as deputy from Chihuahua again, though Gómez del Campo later represented districts in Durango and Yucatán. In 1886 Díaz pointedly rejected Guerrero's nomination for reelection, allowing him to serve only as a suplente from 1888 to 1892.[38]

Luis Huller and His U.S. and British Successors: The International Company of Mexico, the Mexican Land Colonization Company, and the Land Company of Chiapas, Mexico, Ltd.

The name of the German immigrant and resident of Sonora, Luis Huller, never fails to be associated with the massive public land transfers of the Díaz era, invariably in connection with the spectacular size of his company's compensation in Baja California—5.3 million hectares,[39] or one-third the entire peninsula. Huller's participation in the survey company business merits attention for another reason, however. Although he was the direct recipient of only one survey contract in the six-state study area, Huller initiated the most aggressive single campaign to dominate the business. Acquiring nine survey contracts that had originally been let to others, he passed eight of them on to the International Company, which bought an additional concession on its own. The Mexican Land Colonization Company finally held all nine of International's.

Huller's connection with the International Company is not entirely clear; apparently the company was organized after he acquired his first concessions, as a way to raise capital to exploit them. After acquiring his first concession from the Mexican government, he raised $?80,000 from Edgar T. Wells of Hartford, Connecticut, and some New York businessmen,[40] which he used to leverage $500,000 from Drexel, Morgan & Company In March 1888 the International Company's board of directors included Huller; Wells, the president; Richard A. Elmer, a former assistant U.S. postmaster general and president of the American Surety Company; Maj. George Sisson, a San Francisco civil engineer; and William Hamersley, a Hartford lawyer. The company, organized in Hartford and with its head office in New York, already held title to 17 million acres (6.9 million hectares) in Baja California and 4.5 million acres (1.8 million hectares) of land in three other states, according to a prospectus published in London touting the sale of US$3 million in secured bonds in March 1888. The bonds were to raise money to develop the land and certain transportation concessions.[41]

Huller transferred his eight concessions to the International Company, to which land titles were being issued for work done by Huller as early as

December 1887. The company soon found itself unable to raise enough capital to manage its debt, however, and turned to British investors to rescue it. In April 1889, International's stockholders voted to sell the firm's assets and liquidate the company. Negotiations with British capitalists were soon underway, and by July a new English company held all of International's concessions.[42]

The Mexican Land Colonization Company, with headquarters in London, was chartered in Mexico in July with a capital of 2 million pounds sterling, divided into 200,000 shares at 10 pounds each. The subscribers were Joseph White Todd, George Cadell, Charles Morrison, Member of Parliament William Cuthbert Quilter, Alexander Henderson, Henry William Henderson, and Edward G. Parish, all of London. Besides the International Company, the firm also acquired the property of the following subsidiary companies of International: the Yaqui River Mining Company in New York and the Mexican Phosphate & Sulphur Company in California. It also acquired the following Connecticut-based firms: the Mexican International Pacific & Gulf of California Steamship Company, the Mexican Pacific Pier & Warehouse Company, the Peninsular Railway Company of Lower California, the Chiapas Railway Company, and the Gulf of Mexico Guano Company.[43]

In return for all the concessions, land, and other property of the International Company, the latter's shareholders received one share in the Mexican Land Colonization Company for each share they held in International. As of July 1889, International had acquired the concessions and properties itemized in Tables 17 and 18. The English firm also owned four steamships, a hotel in Ensenada, the property of the Yaqui River Mining Company, 200,000 of the 250,000 shares in the Mexican Phosphate & Sulphur Company, and a telegraph line between Ensenada and San Diego. President Díaz approved the transfers, provided that no foreign government would ever hold shares in the new company.[44]

Luis Huller presumably received shares in the new company in exchange for his holdings in the old one, but the entrepreneur's effort to corner a major portion of the survey company business may not have paid the dividends he expected. He died in November 1891 owing his lawyer, Manuel Sánchez Marmol, $10,500 for legal services, which the lawyer had to go to court to collect from Huller's estate in 1894. Huller's son Luis became a Mexican citizen; in 1928 he was still living in Mexico City when his request to see Fomento's records of his father's concessions was rejected on the grounds they had been transferred to the English company.[45]

In 1906 the Mexican Land Colonization Company organized a new company to take over operations in Chiapas. Established with a nominal

Table 17 Concessions Held by the International Company in July 1889

Original Holder	Description	Date
Telésforo García et al.	Survey in B.C.	4/17/83
Luis Huller et al.	Survey, land purchase, colonization in B.C.	7/21/84
Adolphe Bulle	Survey, land purchase colonization in B.C. and Sonora	6/23/84
Telésforo García et al.	Survey in Sonora	2/6/83
Telésforo García et al.	Survey in Sinaloa	1/17/83
Telésforo García et al.	Survey in Sinaloa	6/14/83
Ignacio Gómez del Campo	Survey in Sonora	7/20/83
C. Quaglia and L. García Teruel	Survey in Sonora	12/18/85
Luis Huller et al.	Rights under Quaglia, Teruel contract	6/8/86
Eduardo Valdéz et al.	Survey in Sonora, Sinaloa	9/28/85
Andrés Gutt	Survey in Chiapas	6/4/86
Luis Huller	Land purchase and colonization in Chiapas	11/18/86
Luis Huller	Colonization on Pacific Islands	6/4/87
Luis Huller	Survey, purchase, and colonization of Isla Socorro	6/20/85
Luis Huller	Railroad construction in B.C., Chihuahua and Sonora	5/10/87
Manuel Saavedra	Railroad in Chiapas	4/10/83
Luis Huller et al.	Pacific steamship line	4/16/86
International Co.	Mail between San Diego, CA, and Ensenada, B.C.	5/28/87
Luis Huller	"Productos naturales"—transferred to Mexican Phosphate & Sulphur Co.	4/25/87

Source: ATN. CHIAP 1.71/22, pp. 83–86.

capital of 350,000 pounds sterling, it was called the Land Company of Chiapas, Mexico, Ltd. In return for its assets, the Mexican Land Colonization Company received 30,000 pounds sterling worth of shares in the Chiapas company, plus 170,000 pounds sterling in cash. When the company's lawyer, Emilio Velasco, asked President Díaz to approve the transfer (which he did), Velasco said he hoped the organization of the new company would improve the outlook for "the Mexican land business overseas, which is so discredited today." And because the stockholders—who included S. Pearson & Son, Ltd., the oil company—were such important figures in English finance, "it especially merits the protection of

Table 18 Land Held by the International Company in
July 1889

State	Description	Area (ha)
Baja California[a]	Bulle concession	702,268
Baja California[a]	García concession	1,795,719
Baja California[a]	Huller concession	3,591,438
Sonora	Bought from Telésforo García et al.	245,554
Sonora	Bought from govt between Yaqui and Mayo rivers	43,900
Chiapas	Gutt and Huller concessions	139,955
Isla Cedros	Huller concession	34,217
Isla Socorro	Huller concession	12,636
Total holdings		6,565,687

Source: ATN. CHIAP 1.71/22, pp. 76–86.
[a]Includes some land bought from individuals.

the Government," Velasco added.[46] The eighty-five stockholders, with their shares in the company, are listed in Table 19.

Luis Martínez de Castro and the Sinaloa Land Company

Luis Martínez de Castro, a member of one of the wealthiest and most powerful families in Sinaloa, nevertheless joined hands with U.S. capitalists to exploit his survey contract, the only concession still active in the last decade of Díaz's rule. The contractor's brother Mariano was a senator and governor of Sinaloa in the 1880s and 1890s.

Before acquiring his first survey concession in 1899, Martínez de Castro represented Becerra Hermanos in its survey work in the late 1880s, and was accused by a disgruntled landholder of being a partner in the company, as well. He did invest in Sandoval and Pacheco's Sinaloa concession in 1890, for which he also acted as general administrator and legal representative.[47]

Martínez de Castro acquired his own contract to survey, arrange composiciones, and colonize in Sinaloa and Chiapas in May 1899. He found the colonization obligation too onerous, however, particularly the requirement that at least 75 percent of the families be Europeans. He begged Fomento to drop the colonization clause because it was too hard to attract colonists to Mexico. The government relented, replacing the

Table 19 Initial Shareholders, Land Company
of Chiapas, 1906

Name	Profession	Shares
Madame Coen Albites		1
Baker & Sturdy	Stockbrokers	400
John Hubert W. Baly		1,350
Thomas Bartlett		1,000
Arthur Battiscombe		100
Thomas Bolter		1,800
William W. Brabner		1,350
Charles James Brown		1,760
Harry Brown		250
James Cumm Brown		100
Brown, Shipley & Co.	Merchants	5,000
Walter Burlinson		1,640
Henry Walton Burnside		2,500
George Richard Cable		200
Luis Camacho		250
Geo. Marshall Campbell	Accountant	2,602
Joséph R. Carter		10,736
Robert Cooper	Accountant	5,000
James Thompson Crier		400
Francis Nathaniel Downey		13,000
James Fair		100
Frederick John Gordon	Accountant	3,000
Harold E. P. Gordon		5,300
Archibald Campbell Gray		200
Greenwood & Co.		15,436
Thomas Greenwood		100
Edward Griggs		3,000
Arnold Henderson		1,000
Bernard Henderson		2,000
Lady Jane E. Henderson		2,500
A. Puleston Henderson		5,000
Sir Alexander Henderson[a]	MP	5,000
Brodie Haldane Henderson	Engineer	800
William Henderson		5,000
Hy. William Henderson[a]		20,600
William Frederick Henry		1,800
Norman Gerard Hill		100
Rt. Hon. Charles Hillingdon with Maurice Glenn	Baron	5,000

Table 19—Continued

Name	Profession	Shares
Sir Thomas N. Holdich		250
George John Ingram	Stockbroker	400
Kennedy & Robertson		800
Cecil Humphrey Lambert		500
Thomas Francis Lardelli	Stockbroker	100
Alex Graham Low		250
Marnham & Co.		3,000
John Barry Marwood		100
Inglis Wilfred Mason		7,700
William F. Meres	Bengal civil service, rtd.	1,000
Ernest Allen Millar		100
Percy Bradley Miller		400
Harold James Milledge		2,500
Charles Morrison[a]		5,000
John Noble		100
Thomas Noble		100
Frederick Harold Payne		11,200
Henry Herbert Peet	Munster House	25
Pearson & Son Ltd.	Engineers	5,000
Percy Cuthbert Quilter		14,500
Sir William Cuthbert Quilter[a]		19,000
Charles Foyle Randolph		1,000
River Plate & General Investment Co. Ltd.		1,000
River Plate Trust Loan & Agency Co. Ltd.		1,000
Alexander Nelson Rochfort	Col., R.A.	50
John Henry Roe		2,000
Ernest Ruffer	Banker	1,000
Albert Edward Saunders		1,500
E. Clarke Schomberg		50
Sir Buchanan Scott		250
E. De Grave Sells		100
William Henry Shenstone	Mg. Dir.	100
Michael Snell	Stockbroker	500
Swiss Bankverein		2,000
Robert White Thrupp		100
Joseph White Todd[a]		500
Harry Oswald Tubby		200
Henry Seymour Trower		2,000
George R. T. Upton		3,750
Samuel Watkinson		5,000
Francis Weldon		1,000

Table 19—Continued

Name	Profession	Shares
Mo. Robert Whirter		1,000
Peter Williams		500
Leslie Wilson	Stockbroker	1,000
John Alfred Wood		500
Sydney Young		500

Source: ATN. CHIAP 1.71/24, "Land Co. of Chiapas, List of Stockholders."

^aOriginal subscribers to the Mexican Land Colonization Company, this firm's predecessor, in 1889. Five of the original seven invested in the new company.

contract in January 1901 with one that omitted all references to colonization. The contractor's authorization to survey in Chiapas was nullified in 1905 (for reasons not stated in the record), after he had already surveyed 75,000 hectares in that state. He never lived to obtain any of the Chiapas land because the survey was opposed by Bulnes Hermanos; see chapter 6 for the post revolutionary disposition of the Chiapas land.[48]

After Martínez de Castro had already been given some 185,000 hectares in compensation for his surveys in Sinaloa, he brought in two Los Angeles men to share that concession in January 1905. Albert D. Wallace and Robert P. Probasco paid him $25,000 and agreed to cover all the costs of surveys conducted in the previous ten months whose entitlement was still pending, provided the costs did not exceed $20,000. In return, titles to all the land surveyed in that period would be issued to the newly formed Sinaloa Land Company. The corporation would be Mexican, with a capital stock of not less than $100,000, of which one-eighth would be held by Martínez de Castro. The Americans promised to spend at least US$50,000 for future surveys and land purchases in Sinaloa, and had a one-year option to buy, at $2 per hectare, land that Martínez acquired under the concession before March 1, 1904.[49]

Control of the new company was bought by Wallace, who held 85.5 percent of the shares and became the president. Probasco took just 1 percent. Both Americans described themselves as real estate agents, and within three years control of the firm had been acquired by a Los Angeles engineer, Nelson Rhoades, Jr., who held 93 percent of the company. Rhoades acted as the Sinaloa Land Company's assistant general manager as early as 1906, and by 1911 he described himself as the firm's vice president and general manager. He wrote on the letterhead of the Culiacán plant of the Almada Sugar Refineries Company of New York, but his connection with the Almada works (see chapter 4) was not clear.[50]

Luis Martínez de Castro died in Mexico City in 1920. His widow, Mercedes Esqueda de Martínez de Castro, moved to Los Angeles, where she filed a will in 1927 by which she distributed ten houses in Culiacán and two in Mexico City, as well as several lots in the Sinaloa countryside, to her children and grandchildren.[51] The Sinaloa Land Company was still in operation as late as 1932 and still under the same ownership.[52]

Gen. Francisco Olivares

Among the contractors in the six-state study area, General Olivares received the second-highest total of land in compensation for surveys. Yet he is one of the most obscure of all the contractors. Correspondence of 1885 in the Díaz archives indicates that the surveyor was on close terms with the president, who told him that he was "one of my most esteemed friends." Olivares along with another Sonoran surveyor, Manuel Peniche, founded the Banco de Nuevo León. Olivares's partner in the exploitation of his two Sonora survey contracts was a foreigner, John C. Beatty. The general's request for permission to cede to Beatty his title to one-third of the 3,006,370 hectares he surveyed in the district of Altar—1,002,123 hectares along the U.S. border and the Gulf of California—was granted by Díaz in 1890, a month after the title was issued.[53]

Plutarco Ornelas

Ornelas was the Mexican consul in San Antonio, Texas, when he received his contract to survey in Sonora and Durango. He never did any work in Durango but did survey 60,853 hectares in Sonora's Arizpe district. Macmanus e Hijos, the company's general manager, invested in the concession, along with Antonio Asúnsolo and Jesús González Treviño. At Ornelas's request his title was issued directly to Macmanus e Hijos.[54]

Manuel Peniche

Within two months of receiving two contracts to survey in Sonora, Manuel Peniche transferred them to a resident of Chicago, John A. Kruze, with whom he then established the Compañía de Tierras de Sonora. The other investors in the new company were Samuel Brannan of Nogales, Arizona, and Richard E. Travers, E. J. Horvan, L. H. Ralston, and H. A.

Brown, all of Chicago. The capital stock was five thousand pesos; Peniche and Brannan held one-fourth of the shares, while the rest was divided among the others in the proportion on which they were to decide. Kruze was to manage the firm, while Peniche would represent it before the government.[55]

Carlos Quaglia and Luis García Teruel

A survey contract unlike any other was obtained by Quaglia and García Teruel. Quaglia, a Oaxacan like President Díaz, had fought against the French, and in Díaz's successful 1876 rebellion had fought at the side of General Pacheco, the Fomento secretary. He later served as governor of Morelos and as a senator. During the González administration, García Teruel was awarded contracts to produce uniforms for the Cuerpos Rurales. Both men had already acquired all the survey documents and any rights to land remaining under the Mexican government's contract with J. B. Jecker, Torre y Compañía, which had been hired in 1857 to survey public lands in Sonora, Baja California, and the Isthmus of Tehuantepec. Quaglia and García Teruel traded stock they held in the Ferrocarril Interoceánico de Acapulco, Morelos, México, Irolo y Veracruz worth $150,000 to get the rights to the concession from a partnership made up of Ignacio Altamira, Francisco Búlman, and the defunct company's creditors. Then Quaglia and García Teruel signed an agreement with Fomento in December 1885, under which they ceded more than 175 maps, field books, and other documents associated with the Jecker concession, in exchange for one-fifth of any surveyed public land that appeared in the documents. Of the one-fifth, each of the two men was to receive 17 percent; the rest was to be divided thus among their partners in the enterprise: Pablo Macedo, 34 percent; Leonardo Fortuño, 15 percent; and Carlos David de Gheest, 17 percent.[56]

Fomento then hired Luis Huller to identify and take possession of the remaining four-fifths of the land surveyed by Jecker, for which in compensation Huller would get 5 percent. The five investors in the Quaglia–García Teruel deal had already turned over most of their rights to Huller, each reserving for himself just 8⅓ percent of the land in Sonora and Baja California. Huller in turn transferred these concessions to the International Company and its successor, the Mexican Land Colonization Company. Finally, in 1891 and 1892, four of the five Quaglia–García Teruel investors ceded their 8⅓ percent shares, in the Sonoran district of Altar only, to Manuel Martínez del Río, while the Mexican Land Colonization Company sold him the 5 percent (for Altar only) originally due Huller, giving Martínez del Río a total of 38⅓ percent of the baldío surveyed in

Altar. In exchange for the latter, Martínez del Río promised to pay the company fourteen centavos for each hectare of land he would ultimately be titled by the government under the contract. He conducted the survey in 1891 and reported finding 1.4 million hectares of baldío in Altar. Of the 38⅓ percent he claimed he was due (i.e., 546,192 hectares), he received title to 226,876 hectares. Following litigation in which the Mexican Land Colonization Company successfully argued that Martínez del Río had failed to pay for the land as required by his contract, the title was abrogated and the land turned over to the company instead, which gave it back to the government in 1904 in a swap for property in Chiapas.[57]

Ignacio Sandoval

Sandoval's name was associated with two separate concessions, one for Chihuahua issued in December 1886, and another for both Chihuahua and Sinaloa issued in 1890, which was soon nullified by the Díaz government because of complaints from landholders. (See chapter 3.) Sandoval held the second concession with Enrique Pacheco, and Luis Martínez de Castro was hired to manage the surveys in return for 25 percent of the profits.[58]

Sandoval's concession for Chihuahua alone was operated by Ignacio Sandoval y Compañía, organized by Sandoval, Concepción Sánchez, Andrés Salazar, Guadalupe Gutiérrez, Mariano Lachica, and Ponciano Falomir. All but Sandoval and Falomir, who were residents of Ciudad Chihuahua, resided in San Juan Nepomuceno in Cantón Mina. Sandoval contributed two-thirds of the capital stock of four thousand pesos, and the other shareholders divided the remaining one-third. Of the land to be given the company in compensation, two-thirds would go to Sandoval and one-fifth of the remainder to each of the other five men.

The stockholders must have been landowners in the area to be surveyed, because the company charter gave each of them the right to receive half "of the portion that corresponds to them of the lots adjoining the property they possess or where they may designate." If any land was sold by the company, the product would be divided among the owners according to the above formula.[59]

Jesús Valenzuela

Brothers Jesús and José Valenzuela of Chihuahua obtained a concession to survey in that state that yielded the highest total of land given in compensation, 3,001,551 hectares. They were also active as managers, agents, or investors in surveys in Durango, Sonora, and Sinaloa.

The Chihuahua concession that was generally identified with Jesús was actually issued to six men in December 1882: himself, the ubiquitous Telésforo García, Felipe Arellano, Enrique G. MacKintosh, Dámaso Sánchez, and Ignacio Fernández. A few months later this group also acquired the survey contract awarded to Jesús Díaz González (only two weeks after Fomento gave it to him).[60] Together, these two concessions authorized them to survey in fifteen of the state's twenty cantónes, the other five having been turned over to Gómez del Campo and Guerrero, and to Ignacio Sandoval.

Jesús Valenzuela appeared to do very little in connection with the survey, which was managed by his brother José, in whose name most of the company's business was handled. Jesús, "a rich and famous man," was best known as a poet and as Chihuahua's representative in the Cámara de Diputados from 1880 to 1902; he is remembered today as the founder, with Amado Nervo, of the *Revista Moderna de México* in 1898. Arellano, Chihuahua's senator from 1884 to 1888, headed the customhouse in Mazatlán; President Díaz was the godfather of his son José.[61]

The six men who held the concession each owned one share, except for Telésforo García, who held two. In May 1885, Jesús transferred to his brother José his rights to any land in the cantónes of Meoqui, Jiménez, Camargo, and Aldama for $1,000. A few months later Arellano quit the company. Because his assets, like those of all the partners, consisted only of surveyed land, he received (in proportion to his participation) 52,680 hectares of land in Cantón Camargo, the only territory for which the company had by then been titled, plus another 52,680 in Cantón Bravos after it was titled.[62]

In the meantime, Manuel Herrera and Celso González had joined the company. Sánchez turned over his share to Telésforo García, and MacKintosh transferred his to Celso González. García quit in 1887. González left the following year, getting as a return on his investment 334,594 hectares in Cantón Meoqui and another 195,985 in Ojinaga. Jesús Valenzuela had by this time also abandoned the partnership, having transferred the remainder of his share to García. By 1888 only Herrera and José Valenzuela remained as partners.[63]

Jesús Valenzuela was recorded as the agent for Telésforo García's surveys in Sinaloa and Sonora, and for Refugio Salazar's in Sonora. His brother was much more active, however. He held three-fifths of Asúnsolo's concession and acted as his agent in Chihuahua. He also handled Rafael Váldez Quevedo's survey in Chihuahua.[64] He acquired his own contract to survey in the directoria of Choix in Sinaloa's Fuerte district in 1891 and also represented his sister Natalia, who had acquired Becerra's right to survey in the same location.

Policarpo Valenzuela

A veteran of the war against the French, in which he attained the rank of colonel, Policarpo Valenzuela became one of the richest men in Tabasco. In 1910, four years before his death, an official biographer reported that, besides owning the railroad from Villa Cárdenas to Río Grijalva, Valenzuela was a shipowner, merchant, president of the Banco de Tabasco, and "partner in a huge number of industrial, agricultural and economic businesses, since the range of his trading operations knows no limit." He cut timber, raised sugarcane, and operated a firm in the state capital that ran five steamships up and down the state's rivers, "the main part of a great fleet that once made up the only comfortable means of internal transportation." Elected governor of Tabasco in 1910, he took office on January 1, 1911, but was forced to step down six months later when the Díaz regime was replaced by the government of Francisco Madero.[65]

Valenzuela served two brief terms (of less than a month each) as interim governor in 1886 and 1887 before acquiring the survey contract awarded to Manuel Sánchez Mármol, the politician-lawyer-journalist who represented the interests of surveyors Bulnes, Méndez Rivas, and Huller. President Díaz did not noticeably favor Valenzuela, turning down his request in 1886 that the $33,750 owed him for the subsidy of the railroad he built be provided in the form of exemptions from export and import duties, which Valenzuela blamed for certain "heavy losses." Díaz not only rejected the request, but reminded Valenzuela that the subsidy only came to $8,000. The president also pigeonholed Sánchez Mármol's request that someone issue Valenzuela's credentials promoting him from colonel to general, a rank to which Díaz himself had named him. Alas, the magnate died a colonel.[66]

Appendix B

Allegations of Land Usurpation in the Six States

All allegations of land usurpation found in Fomento's survey archives are summarized below, with two exceptions: post-1900 surveys in Sinaloa (which are only summarized here), and two cases (that of Bulnes Hmnos. in Tabasco and Becerra in Sinaloa) that are reported in chapter 4. See the introduction to that chapter for the methodology and certain caveats regarding the data that follow. The cases are organized first by state, and then within states by survey company.

CHIHUAHUA
Ignacio Gómez del Campo and Ramón Guerrero

Eleven "indígenas tarahumares de Sisoguichic" in Cantón Abasolo saw the notice in the state's *Periódico Oficial* giving property holders whose land adjoined that being surveyed by Ignacio Gómez del Campo and Ramón Guerrero thirty days to object to the survey. In response to the notice the Indians petitioned the federal district judge in Chihuahua in June 1883, informing him that the seizure of their lands would impoverish them, and presenting titles of ownership issued by President Juárez. They made no specific accusation against the company, but seemed motivated instead by fear of the consequences of a survey. The Indians were so poor that the judge granted their request to affix five-centavo stamps to the documents instead of the more expensive ones usually required. The judge acted on no other evidence of their poverty than their appearance, "since some of those present are not even wearing shirts and all of them are dressed with the greatest simplicity," the court reported.

Six months later the company's representative, Guillermo Urrutia, assured the judge that the Indians' "posesiones" would be respected, whether or not protected by title. Urrutia said the company was unaware of the Indians' claims, since they had not appeared during the survey itself, and expressed hope that the case would not be delayed any longer.[1]

In the adjoining cantón of Balleza, the company's survey provoked fear of an uprising by the Tarahumara Indians. When Fomento asked the company for an explanation, it claimed in March 1884 that the alarm was unfounded. Far from opposing the survey, the Indians were permitting it in spite of advice to the contrary by certain whites who, acting in the name of the Indians, "have usurped large and good-quality portions of national land." Individuals from San Pablo and other places, the company continued, have for years been squatting on as much as forty or more sitios of ganado mayor each, or some 70,240 hectares. These people, who long assumed that nobody else would want the land, never bothered to submit land claims to Fomento (i.e., under the 1863 law). And now that the land is worth "a good sum, they don't want to pay it." These people, the survey company argued, were trying to stir up the Tarahumaras, but the latter were not paying attention to them. The Tarahumaras "agree that the ejidos of their pueblos and those small properties that they possess and sow each year should be respected, a position to which as you know the company defers, the matter being settled according to the law and to various supreme dispositions," the company stated in a letter to Fomento.[2]

Nevertheless, three months later the jefe político of Hidalgo district reported to the governor of Chihuahua that the survey has "greatly alarmed the Tarahumaras, and grave events may be feared if the survey is carried out. A conflict with the highland Tarahumara is to be feared because being semi-savage people, great and costly sacrifices would be necessary to subdue them."[3]

In the event of war with these tribes, the jefe político continued, communications with the mines would be cut off because the tribes occupy nearly all of the Sierra. This would deprive the state "of its exploitation and of the advantages that are provided by it." The uprising of "tribes so powerful in number would be the complete ruin of the State" just to satisfy the private interests of a survey company (because the land involved is unfit for colonization anyway).[4]

Fomento must have believed the company, because a request for an extension of its survey contract was readily granted by President Díaz in July 1884. The following year a Fomento functionary reviewing the judicial record of the survey sent by the local federal judge noted that the survey was conducted "with all the legal formalities, until it was completely finished, without there arising any circumstance or obstacle worthy of note."[5]

Balleza soon became the scene of another complaint involving the company. Gen. Carlos Pacheco, the secretary of Fomento, sharply criticized Gómez del Campo and Guerrero in 1886 for claiming as baldío

1,798 hectares owned by Pedro Carrillo, and then sold by the company to Faustino Terrazas. When Carrillo complained to Fomento, his claim to the land was checked and ultimately deemed valid, Pacheco told the company, adding that he would see to it that the federal judge in Chihuahua returned the land to Carrillo.[6] General Pacheco, who continued to hold his post as governor of Chihuahua by running the state through subordinates there,[7] took the occasion to blister dishonest survey companies who, just to avoid a contested proceeding, promised landholders they would respect their property and then informed Fomento the land was baldío. The general told the company:

> The Government already has a variety of information regarding the abuses that the Agents of the Survey Companies have committed, enclosing private properties on the maps which constitute the objective of their operations, by means of which they shock the good faith of certain adjoining property owners, or they exploit their ignorance, offering to respect their country estates, persuaded of the soundness of the titles that apply to them, with the aim of avoiding contested proceedings . . . committing the government to approving, as national land, territory that is far from having that character.

Such was the case with Pedro Carrillo, the general continued, virtually accusing the company of fraud. The survey map failed to show the property "so that it would be confused with the baldío that was adjudicated to the Company, at your request," and which the company later sold to Faustino Terrazas. General Pacheco sent the matter to the federal district court so the land could be returned to Carrillo.[8]

At about the same time Ramón Guerrero, Gómez del Campo's partner in the company, clashed with villagers in Balleza when he tried to sell land he had acquired in the survey. In the process of selling part of the newly surveyed land to three buyers, he apparently usurped land claimed by the pueblo of Huejotitán. The vecinos sought relief before a local judge in 1887, but by then the sale had been completed. According to a representative of the vecinos writing to Fomento in 1889, the vecinos were later "assaulted in the homes of their families and threatened with weapons" by the surveyor hired by the buyers, by the judge who authorized the new survey, and by the buyers themselves "and others with repeater rifles in their hands, a total of fourteen men." When the vecinos asked the ministry to restore their land, a Fomento functionary studied the case and found that the pueblo's titles to five and one-quarter sitios of ganado mayor (9,219 hectares) and one sitio of ganado menor (780 hectares) were issued in 1749. The survey company's map omitted any reference to this titled land, except for one sitio (1,756 hectares) of ejido. Fomento decided that the titles were valid and forwarded them to the appropriate court, "so that the justice which the complainants are deserving of may be imparted to them." The ministry gave the company three choices: negotiate an agreement privately with the vecinos, allow a court to decide the matter, or simply return all the pueblo's lands. If the company took the latter course, Fomento offered to replace the land with some of the government's two-thirds resulting from that survey, but not before reducing the one-third the company got in compensation, "because it does not belong to it since it surveyed a private property."[9]

In his defense, Ramón Guerrero argued that all the holders of adjoining land had signed the survey document. After he quit the company, Guerrero explained, he sold his share of land to two men in 1886. Then "various inhabitants" of Huejotitán, who occupied some of the sold land "without legal title," refused to leave it. Guerrero said that during the colonial period the Indians' cacique in Huejotitán made an agreement with the government to buy five and one-half sitios (9,658 hectares) to establish a pueblo to be called Misión, in exchange for a certain sum of money. The deed was issued but the cacique was unable to pay for the land within the period specified. Thus, Guerrero argued, the land reverted to government property and has been occupied "without real right." Moreover, the government itself later sold part of the five and one-half sitios in question, and the vecinos never alleged any right to that part. Anyway, he added, the Mexican constitution has abolished "community land."[10] As will be seen, it was not unusual for surveyors and the government to respect community or ejido land attached to a particular town, even though the constitution required that it be privatized.

Guerrero appealed to Gov. Lauro Carrillo to use his good offices to resolve the conflict with the vecinos, whose lawyer had so far been unwilling to settle. Guerrero said he was willing to give land back to anyone who could either produce a legitimate title or prove longtime residence, as long as Fomento compensated him. Governor Carrillo urged Fomento to appoint a surveyor to determine the ownership of the lots in dispute and to compensate Guerrero for those he would lose, in hopes that this arrangement would satisfy all the parties.[11]

José Antonio Rodríguez, who represented Huejotitán in its dispute with Ramón Guerrero, complained separately in 1889 that land he owned was also being sold by Guerrero, though he admitted that the company had cited him to appear for the survey and present his title, and that he had failed to do so because he was away from home at the time. A Fomento official who investigated the complaint found that Rodríguez's title was legitimate but that he had only himself to blame for the usurpation, because he had failed to present his title when told to do so. Nevertheless, the government sternly informed Guerrero that the land issued in compensation to his survey company wrongly included Rodríguez's lot. As a result, Guerrero was told, President Díaz had decided that the land should be returned to the rightful owner, though the company could choose its third from the two-thirds reserved for the government (presumably because Rodríguez shared the blame).[12]

Just as the vecinos of Huejotitán were mobilizing themselves, sixteen vecinos who held two ranchos in the neighboring municipio of Olivos complained to Fomento about the surveyor who was checking the limits of the land that Guerrero had just sold. They claimed that the surveyor was enlarging, at the expense of the property holders, Guerrero's legitimate 597-hectare strip of baldío between two pieces of property, "as they say happened with Huejotitán," according to a Fomento functionary's memo. His recommendation, that the vecinos be told that only the courts could handle this problem, was approved by a superior, who added this penciled note to the memo: "that the complainants may always have recourse to the courts and to the Government in the event they trespass into their boundaries." Fomento told the petitioners that the survey company's map did show their property and that the survey by the new buyers should therefore not affect their land, but if it did they ought to appeal to the court and to the governor.[13]

The company's survey of Cantón Galeana, at the opposite end of the state along the Texas border, provoked several complaints among other land developers. The survey, conducted in 1883–84, identified 3,026,110 hectares of baldío in that cantón—the largest single survey conducted by a private contractor in Mexico.

In February 1887 Fomento ordered Gómez del Campo to appear at the ministry to answer for the "grave irregularities" of his Galeana survey, discovered when the ministry studied the maps he submitted and when "various public land companies" presented claims. The purpose of the meeting would be to negotiate a fair settlement with the latter companies. Gómez del Campo replied that he was not entirely to blame, because he was a partner in the company with two others—Ramón Guerrero and Señor Macmanus. His request for a twenty-day extension was granted so that he could go to Chihuahua and investigate; a meeting was later scheduled for March 26.[14]

In 1890 four complainants were sold, in compensation for the survey errors, a total of 754,912 hectares at twenty-four centavos per hectare, payable in government bonds. They claimed possession of what the government had received as its two-thirds share of the survey, but because that land was already disposed of, they were sold in compensation land located in three cantónes—Jiménez (294,814 hectares), Degollado (424,073 hectares), and Galeana (36,025). In the titles to the land, signed by President Díaz, he cited the "respect that he has always had for private property," even though in this case no one protested during the survey itself, leaving the government to assume at the time that the land was baldío. Gómez del Campo, as the agent of the company (defunct by the time this deal was made in November 1889), agreed to pay the four claimants $32,843 in government bonds to compensate for that part of the one-third the company had wrongfully acquired as a result of the survey. Two of the buyers were the heirs of José Ignacio Ronquillo and Félix Gameros. The other two—Pedro and Manuela Olivares, and Domingo Leguinazabal—bought 316,000 hectares, of which they requested that 158,000 hectares be titled directly to José Valenzuela. The Olivares also sold 87,800 hectares of their land to Lauro Carrillo, the governor of Chihuahua when the agreement was signed.[15]

Antonio Asúnsolo

Two complaints surfaced against this company's survey of part of southwestern Chihuahua. In 1897 the jefe político of Arteagas district informed Fomento that Asúnsolo's company had invaded part of the ejido of the village of Chínipas in the process of attempting to acquire legal possession of some land it had surveyed. No response from Fomento was recorded in the file.[16]

In another case the vecinos of the *mineral* (mining town) of Batosegachic in Arteaga district complained in 1891 that the Asúnsolo company had failed to respect their ejidos. They said they were entitled to 1,755 hectares under the law of the state and also wanted an additional 145 hectares. A Fomento memo, citing Asúnsolo's record of the survey of Cantón Matamoros,[17] recommended issuing Batosegachic a title to the land it wanted. A month later, however, it was discovered that Batosegachic was merely a *barrio* (district) of the municipality of Guazapares and therefore was not entitled to ejidos. A recommendation was made that the land in question be part of the two-thirds due to the government, and that the latter sell the petitioners the land they wanted at $0.30 per hectare, minus a 25 percent discount (as provided by the laws of prescription). The secretary of Fomento directed that a letter be sent to the governor of Chihuahua, transcribing the petition and noting that the place was not entitled to ejidos. The letter was sent, but here the file ends.[18]

No evidence exists that the company ever received title to any land it surveyed in Cantón Matamoros. The only titles that turned up for survey compensation were for land in former cantónes Andrés del Río and

Allende. Substate boundaries were difficult to pin down, however, and the title for 246,311 hectares in former cantón Andrés del Río, issued in 1891,[19] could have included land in the adjacent cantón of Matamoros. Moreover, the complaint from Chínipas cited earlier indicates that the company held title to land in Arteaga (formerly Matamoros).

Ignacio Sandoval and Enrique Pacheco

Signs of rebellion among the Tarahumara people in Cantón Galeana did not surface until after land first acquired by Sandoval and Pacheco was later being subdivided and sold. Mateo Carrillo, identifying himself as the "headman of the lowland Tarahumara," told Gov. Lauro Carrillo on May 5, 1891, that he represented the pueblos of the Baja Tarahumara. Three years ago he had received "an infinite number of complaints from the greater part of the Pueblos" over the plundering of their lands, some of which were titled by Benito Juárez, and which were now being resold after they had been bought by the survey company owned by Enrique Pacheco and Ignacio Sandoval. Fourteen pueblos were under his command, the Indian leader told the governor, and "many *rancherías*" with five thousand to six thousand inhabitants whose lands had been reduced. Pointing out that he had unsuccessfully appealed for justice at Guadalupe y Calvo, the *cabecera* (capital) of Distrito Mina, he warned the governor that there were now signs of rebellion among the pueblos, and that he would not be responsible if an uprising occurred, because he had tried to calm them. Eleven days after the Indian leader wrote the letter, Gov. Lauro Carrillo sent a copy of the complaint to Fomento and noted that he had told the Tarahumaras that all their titled lands would be respected, and that he would support their claims to titled land, but that he would punish any Indians who rebelled.[20]

In response Fomento told Sandoval and Pacheco on July 20 that President Díaz had suspended all of their surveys in Chihuahua because of the Tarahumara complaints. The company's appeal for a reversal of the suspension, submitted in part because the holders of the affected land had given their approval of the survey when it was conducted a year earlier, was rejected by Fomento in November, despite Sandoval's warning about the "grave damage to proprietors who have honored me." He pointed out that the complaints began three years ago, two years before the land was surveyed, indicating that the survey itself had not touched off the Indians' complaints.[21]

The suspension may have been annulled later, however. Evidence that the company eventually may have received some land as a result of the survey surfaced twenty-nine years later when nine "farmers and heads of family living in the rancho of Santa Rosalia, municipio of Guadaluple y Calvo," having noticed the decree signed by Venustiano Carranza on April 20, 1920, that declared the survey by Enrique Pacheco and Ignacio Sandoval null, wrote to the Secretaría de Agricultura y Fomento to complain that the company had "despoiled" their land. They had made "repeated protests that in those times and before those authorities" were to no avail. The company sold the land to a Señor Morrison, who sold it to Conrad Loya y Mascareñas; it was now held by Señora Leonila Loya de García. The nine complainants then requested title to the two-sitio (3,512-hectare) rancho Rosalia so they could divide it among themselves. The revolutionary government was not sympathetic; the secretaría replied that the matter was out of its hands and would have to be settled in court.[22]

Sandoval alone had been awarded a separate survey contract four years before he and Pacheco acquired theirs. Sandoval's survey of Cantón Mina, however, did not appear to provoke opposition until 1905, thirteen years after he had received two titles for a total of 633,528 hectares of land in compensation. Three hundred Tarahumara residents of the pueblo of Redondeado in the municipio of Guadalupe y Calvo complained that Sandoval was selling 30,000 hectares, which included land they used for grazing. Fomento noted that the 30,000 hectares were probably part of Sandoval's one-third. The ministry informed the governor of the complaint, and a few months later the residents applied for free land under a law that authorized land grants to "labradores pobres."[23]

Rodolfo Valdéz Quevedo

Cantón Galeana was also the site of a dispute involving Rodolfo Valdéz Quevedo's company. When it surveyed a sixty five-sitio section in 1888, the owners of the haciendas of El Alamo, Bocas, and Vallecillo opposed the survey, accusing the company of designating their land baldío. But their opposition was never *formalizado* (filed in court). Ten years later José Valenzuela, describing himself as the *apoderado* (agent) of the company, got the federal district judge in Ciudad Juárez to certify that there was never any "formalizado oposición" to the survey, then asked Fomento for the company's title to the one-third it was due under the survey. The file ends with Fomento's promise to study the matter, but there is no evidence a title was ever issued in compensation for the survey.[24]

Jesús Valenzuela and Company

In March 1885 Pedro Zuloaga, represented by Enrique C. Creel, opposed Valenzuela's survey of Abasolo district, claiming that the company had included his already-titled land as baldío. Rafael Loya y Hermanos, owners of the Hacienda de la Cienequito, filed an identical complaint, as did Nicolás Torres and Mariano Ochoa. A fifth party opposed the survey on the basis of a previous land claim. Valenzuela quickly corrected his maps in response to Zuloaga's complaint and stated that his company did not, in fact, consider Loya y Hermanos' land to be baldío. The Loyas withdrew their opposition. In the other cases, where the land titles had been issued by the state, the company argued that the titles should be invalidated because they had not been recognized by the federal government. By June a sixth landowner had joined the protesters. In October the district judge granted Valenzuela's request for permission to make "some changes" in his maps. Zuloaga dropped his opposition the same day. As for the rest of the plaintiffs, Valenzuela acknowledged the possibility that the land at stake may have been privately held, and the judge granted his request to send to Fomento that part of the file covering land in southern Abasolo not subject to any counterclaim. In the meantime, litigation over the 12,000 hectares still in dispute continued. When Fomento issued the company a title in December to the 138,764 hectares it was due in Abasolo, the 12,000 hectares were pointedly excluded.[25]

At times the land-speculating survey companies were themselves the victims of other land speculators who, sensing a boost in land values once a survey contract was signed, tooks steps to acquire property before the survey company got to it. The Valenzuela company complained to Fomento in 1883 about the opposition it was facing in several parts of the state, not only by local judges but by private property owners and "many other individuals who because of the authorizations [to survey, received by the company] had come to acquire land for speculative purposes." In 1885 "numerosas" oppositions were still on file in the federal district court in Chihuahua, stalling approval of the survey of the western part of Cantón Aldama. A year later the company told Fomento that despite the federal court's decision to reject the protests, the company thought it "prudente" to deduct the disputed area from the total surveyed before requesting title to its one-third. Three claims totaling 22,259 hectares were subtracted, at the company's request, from the 295,390 hectares originally designated as baldío. The same José Horcasitas who had submitted the rather questionable opposition to Valenzuela's survey in Iturbide (discussed later) also opposed this one, but the Valenzuela company successfully resisted it, as well. Fomento found that Horcasitas's title was defective and told him that President Díaz had determined that his complaint was a matter for the courts to decide.[26] Among the Cantón Aldama complaints was one submitted by José María Falomir, who told

Fomento in 1885 that Valenzuela had included his land as baldío. The ministry deemed Falomir's title valid and informed Valenzuela that President Díaz had decreed that Falomir's property would have to be respected because it was legally titled. Within a week Valenzuela replied with assurances that it would be.[27]

Three protests against Valenzuela's survey of Bravos—two of which were based on prior public land claims—were withdrawn in 1885. The Martínez del Río family dropped its opposition after the two sides agreed on a boundary between the family's holdings and the baldío that Valenzuela had found. The other two (one of which was submitted by a group of men that included Enrique Creel and Pedro Olivares) desisted after Valenzuela agreed to recognize their claims.[28]

After the thirty-day notices were published in Valenzuela's survey of Camargo district in March 1884, Manuel Gómez y Luna, Pablo Ochoa, and others opposed the survey on the grounds that they had already filed land claims and thus had prior rights. Nine months later they withdrew their opposition under the condition that the company respect the land they had already claimed; the claimants also promised not to seek any of the newly surveyed land in the area designated as Zone A.[29]

Valenzuela encountered trouble with several landowners in Cantón Degollado in 1885, but all but one withdrew their opposition before Fomento approved the survey. The lone holdout was Luis Terrazas, whose claim was temporarily segregated so that Valenzuela could obtain title to the company's one-third. Settlements were reached with a group of public land claimants who, though they had submitted a claim in 1881, never bothered to survey it. As Valenzuela pointed out, "we could not respect [survey] points that did not exist." Another landholder withdrew after the 63,216 hectares he claimed were dropped from the survey, and Gómez del Campo's survey company dropped its claim that Valenzuela was taking part of the land it had been titled in Cantón Galeana. Valenzuela bitterly attacked Terrazas's continued opposition, which he asserted was baseless and had been submitted merely to delay the survey; Terrazas, he said, did not even present a title to back up his claim to the Hacienda de San Miguel Bavícora. The company eventually agreed to return 37,476 hectares claimed by Terrazas as part of the hacienda; in compensation, Fomento let Valenzuela have an equal area of terreno nacional in Cantón Iturbide.[29]

Two men who identified themselves as "indígenas vecinos" of the pueblo of Cocomórachi in Cantón Guerrero complained to Fomento in 1889 that the Valenzuela company had designated as baldío some land they owned in Degollado, and furthermore had failed to notify them of the survey. The survey company did not identify the land as theirs because their claim to it was still "in process" and they were now seeking Fomento's approval of the claim, which had already been accepted by the federal district court. They had possessed the land "from time immemorial . . . having fought the barbarians in more than fifty years of war that the State sustained with a cruel and bloody [illegible], all facts of public knowledge." The men grazed their cattle on this land, had begun cultivating it, and had homes on it. If they lost it, they said, they would have to emigrate. Nothing further appears in the file after Fomento's letter to Valenzuela requesting a report.[31]

Enrique C. Creel submitted, on behalf of Pedro Zuloaga, a protest against Valenzuela's survey of Cantón Guerrero in 1884. It was soon withdrawn at Creel's request, however, after an unspecified settlement was reached.[32]

Some complaints were clearly untenable. In 1889 the family of Matías and Pascual Escarcega in Guadalupe, Cantón Guerrero, said it had submitted a claim to some public land in 1869, which Valenzuela's company had subsequently classified as public in its survey in 1884. Three years after Valenzuela received title to the company's one-third share the family asked Fomento to let it continue processing its claim, which it had allowed to lie dormant for twenty years. Fomento replied that the company had in fact not counted the land as baldío, and in any case all the public land in that district was already sold.[33]

When the Valenzuela company surveyed Cantón Iturbide in 1885, José Horcasitas refused to present the titles to his Hacienda de Mápula, claiming he was not obligated to do so; he further asserted that the company had no right to survey his property. That the company could scarcely be expected to know where his property was without the titles seemed irrelevant to the complainant. Valenzuela sued Horcasitas for damages for failing to answer the summons to appear with his title at the survey. When the defendant appealed to Fomento to suspend approval of the survey on the grounds it invaded his finca "Hormigas," a government investigation determined that he had few grounds for complaining. When Fomento studied his title it was found to be defective because it had never been approved by the *Audiencia* (institution of the Spanish colonial government), and there was no evidence that the original titleholder had transferred the property to Horcasitas. The latter was informed that President Díaz had determined that his complaint was a matter for the courts to decide, Fomento's usual way of disposing of landholders' complaints that it did not want to handle.[34]

When Jesús Valenzuela's company surveyed the mineral of Santa Eulalia in cantón Iturbide in 1885, the court record indicates that it allowed twenty-five hundred *varas* (yards) from the church door for the town's ejido, "against which the President of Santa Eulalia protested." A law of December 23, 1851, was cited to justify this operation; the date corresponds to no federal land law but probably reflects a state law controlling the size and location of ejidos. In any case, the reason for the protest was not recorded, nor was its outcome.[35]

In Valenzuela's survey of Iturbide district, the company vigorously resisted the efforts of Miguel Salas and two others identified only as Terrazas and Maceyra to retain demasías found on their properties after the survey. In response to Valenzuela's complaint, Fomento first defined demasías as "those properties that surpass the [total] protected by the title, and which, by tradition or any other reason have been taken in possession." The ministry then pointed out that a survey company was permitted to ascertain which lands were actually baldío in those cases where no title, no possession, and no tradition of possession existed. In such a case, the land was "solely baldío" and the company had a perfect right to identify it as such. Fomento's letter apparently did not help Valenzuela very much, because he continued complaining about the trio's claim, which if granted "would make useless the work of the survey companies, and their great expenses without any return of any kind" and "would amount to . . . destroying almost all of the discovery of the national lands," hurting the country as well as the company. Why, Valenzuela demanded, did these three not file their claim before the survey? Three months after the company wrote its first letter on the case, Fomento replied with its final word—President Díaz had decreed that the company had no right to the demasías, and the letter left little doubt that Díaz regarded the land in this case as such.[36]

Four counterclaims were filed in opposition to Valenzuela's survey of Jiménez district in the spring of 1884, including one by Antonio Asúnsolo, himself a survey contractor, who argued that Valenzuela's company had wrongly identified part of the Hacienda de Cañas, which Asúnsolo owned with others, as baldío. One complaint was settled when the plaintiff agreed to pay Valenzuela $2,000 within six months to compensate for any demasías to which the company might have a right. All the others withdrew their opposition after Valenzuela redrew the maps.[37]

SINALOA
The International Company

A Fomento memo in September 1889 noted that in "some sections of minor importance" International's survey of Sinaloa provoked opposition by "the Indians of Abuya and other pueblos." The ministry was waiting for the two sides to settle before acting on the survey.[38]

Mariano Gallego

Mariano Gallego's survey of Fuerte district in 1884 provoked opposition by two separate groups of men who said they had filed public land claims on part of the area being surveyed. One group failed to file the necessary court documents within the thirty days allotted, and the other told the court it had decided to drop its opposition.[39]

Luis Gayou

Nearly a year after Becerra transferred his contract to Luis Gayou, the latter's survey in Mocorito provoked a bitter complaint in August 1889 from Ignacio de la Cueva, who claimed ownership of two parcels whose titles went back nearly two hundred years. De la Cueva wrote to President Díaz, stating that the Becerra company (he was evidently unaware it had changed hands) was owned in part by the governor of Sinaloa, Mariano Martínez de Castro; the latter's brother Luis, deputy to the state legislature; and Manuel Monzón, a local judge. He charged that the company was surveying Sinaloa in "a manner so criminal that a great uproar was now being raised in the State," because its behavior was like that of "the highway bandit ambushing his victim," except that in this case the victim was the state of Sinaloa. Furthermore, "from the moment the company decides to survey a property, it enters and without giving the proprietors or possessors the most [document torn], interference nor any knowledge, it gives orders to open breaches, to suspend the work of the owners of the property and it always has the knack of locating the baldío it proposes precisely inside the possessions that the owners have protected [by title]."[40] Besides, the judges were so "docile that they are not doing any more than what they are ordered to do." De la Cueva suspected that company agents had destroyed his petition to Fomento.[41]

Fomento acted quickly, telling Gayou to forward a report on the complaint and ordering the federal district judge to suspend all work on De la Cueva's fincas and to send the court file of the survey. Gov. Mariano Martínez de Castro, defending himself in explicitly class-conscious terms against De la Cueva's "extremely offensive" petition, called him "a very poor and unhappy man who lacks any kind of property, who may hold in some corporation [comunidad civil] some insignificant right," and besides lacks "the most rudimentary notions of primary instruction." Not only did Becerra no longer hold the contract, but he himself had withdrawn as a partner in the company when he became governor, "as a prudent measure" to avoid "all suspicion of inappropriate influences on my part." The records do not indicate how the conflict was resolved, but no record exists of Gayou having received a title in compensation for any surveys.[42]

Natalia Valenzuela de Saracho

Becerra turned over his right to survey in part of Fuerte district to the only female survey contractor encountered in this study. Natalia Valenzuela de Saracho, the sister of José Valenzuela, provoked two complaints in her survey of the directoria of Choix. One was registered in 1892 by sixty-five-year-old Jesús Islas, who described himself as a *comunero* (joint owner) of San José del Ranchito. He and others had lost the title they held to their land, so they were unable to present it at the time of the survey. Protesting "the abuses committed by the Madame surveyor on my property," Islas declared that "taking advantage of the loss of the title and of the lack of capacity that we comuneros have to defend ourselves, the Madame surveyor gave the order to carry out the survey without giving me notice. Today I am oppressed and despoiled of my fincas and in short of all of my property."[43]

A few months later, Cirilo Pellegand's lawyer protested as well, alleging an invasion of his property. The survey was done "maliciously" because it was conducted while Pellegand was away on business and unable to attend it. Instead of placing the demasías in an unpopulated area, Valenzuela located them within perfectly titled territory, in "places inhabited and cultivated by me over the course of long years of work."[44]

Pellegand filed a complaint in the local court, and nothing further appears on Islas' complaint, but in November 1893 the "señora deslindadora" lodged a complaint of her own against Fomento regulations that protected property owners from surveyors, and appealed for the release of the titles she was due. The requirements that surveyors obtain copies of titles as well as documents from property holders stating whether they renounce the right to buy their own demasías were "almost impossible" to fulfill, she objected, and her surveyors did not make copies of titles when they did the survey. The requirements also placed surveyors at the mercy of the hundreds of property holders with whom they had to deal. "Although the survey operations have a bad reputation in other States, due to the irregularities committed in carrying them out, that has not happened here." Furthermore, landholders here "look with indifference on the survey of the demasías" because of the poor quality of the land and its low value.[45]

Luis Huller

Luis Huller's survey in Culiacán district in the late 1880s not only conflicted with Becerra's but also sparked several complaints from groups of individuals who claimed that the survey threatened their land.

The fiercest resistance came from the *alcaldía* (mayor's office) of Sataya, adjacent to Navolato in the directoria of Altata. This unfortunate village's land was coveted by the governor of Sinaloa, Gen. Francisco Cañedo, as well as by Huller. When Huller's surveyors went to work in December 1886, Governor Cañedo claimed 1,091 hectares that he had bid on in a court-ordered auction. Fifteen individuals "and the rest of the joint owners and possessors" of Sataya protested the survey on the grounds that the land was theirs. They also charged that Governor Cañedo was keeping the titles to the land "in secret," forcing the co-muneros to resort to an old and crudely made sketch of the land on which he was bidding. The sketch proved that their land was actually out-side the limits of Sataya and was therefore baldío, for which they had en-tered a claim that spring. The comuneros claimed that Huller intended to give the land he surveyed to the governor, who was having trouble ac-quiring it as a bidder in court. Neither the surveyor nor the governor was entitled to land of which the comuneros themselves were the "oldest possessors."[46]

Huller argued that their baldío claim of four thousand hectares ex-ceeded the legal limit of twenty five hundred per claim, and that the court's authorization of the survey in 1883 preceded the public land claims entered by the comuneros. Then Governor Cañedo produced a copy of a colonial-era title covering land in Sataya, which was rejected by both the surveyors and the local judge as a probable fake because such important information as the quantity of land and its location was miss-ing where someone had purposely burned away portions of the title. This left the residents of the town as well as the hapless governor no choice but to depend on the "faithful knowledge and understanding" of the survey-ors, as the judge put it.[47]

Other protests against Huller's survey were recorded by the self-described Indians of Abuya in the directoria of Quilá, who claimed in January 1888 that, "abusing our ignorance they declared our titles bad with false promises [and] obtained our consent, which we withdraw."[48] Elsewhere in Culiacán, three other landowners protested separately, claiming that their property had been invaded by Huller.[49]

In response to the complaints the federal prosecutor in Sinaloa filed a motion in federal district court in November 1889 requesting the nulli-fication of Huller's concession, which by then had been transferred to the International Company. (Huller had acquired his right to survey in Sina-loa when he bought two contracts issued to Telésforo García on February 12 and June 14 in 1883, and another originally given to Eduardo Valdéz y Socios on September 28, 1885.) All of the holders of these concessions, but especially Huller, the prosecutor argued, had committed eleven "grave irregularities." Most of them had to do with complaints that the surveyors failed to respect someone's land, or that they had not properly notified landholders whose property bordered the baldío. The judge refused to rule on the motion, forwarding it to Fomento instead, which had still not officially responded as late as 1897.[50]

But in 1892 Fomento had told the federal judge to give Sataya a sitio de ganado mayor for its fundo legal and ejidos, and had furthermore rejected the survey of Culiacán because neither the prosecutor nor the federal judge had yet expressed his approval, and because documents related to the survey of seven lots were missing. The ministry finally decided that "there can be no doubt about the irregularities and omissions that nullify and make valueless the works that are improperly called surveys by the Huller Company."[51] By 1897 none of the opponents of the survey in Culiacán had yet bothered to formally file complaints in court, nor had they taken any steps to perfect their titles. All of them were still holding the same land that Huller had claimed was not titled. Fomento decided to act to prevent "the use or exploitation" of what it nevertheless considered to be public lands by having its agent in Culiacán exercise jurisdiction over the property until it could be sold. There is no evidence that Huller ever received a title for the land that he or his successors claimed to have surveyed in Sinaloa.[52]

Ignacio Sandoval and Enrique Pacheco

Some survey contractors were authorized to arrange composiciones with the holders of land for which no clear title existed. Unlike the *denuncio* (public land claim), composiciones were exclusively the privilege of those who actually possessed the land in question, though without the proper legal documentation. The procedure was sometimes adopted by property owners on whose land a survey company discovered demasías. The landholder paid the company for the survey, and in return the company promised to deliver a title, issued by Fomento, establishing that person's ownership of the demasía. The company paid Fomento the price of the land (presumably included by the company in what it charged the property holder), which was heavily discounted from the biennially fixed prices that claimants of baldío had to pay. The company kept one-third of the price paid for the land. The landholder also had the option of arranging the composición directly with Fomento, in which case the survey company would still get one-third of the revenue from the sale of the land. In any event, a survey company undertaking a composición had a clear incentive to prove the holder's right to the land in question, because if the holder did not meet the legal qualifications for obtaining title to the demasía, the company would be unable to collect the full price it set for delivering the title.

Of the six states in this study, only in Sinaloa were composiciones arranged with any frequency by survey companies. Sandoval and Pacheco's company limited itself exclusively to arranging composiciones, which it did for twenty-five pieces of property. Evidence indicates that titles were actually issued for at least twelve. As we shall see, eight of the twenty-five were abandoned by the company. Several complaints were recorded from individuals who charged that the company had taken their money but reneged on the promise to deliver a title. In October 1891, fifteen months after the survey contract was signed, President Díaz canceled the company's authorization to survey in both Sinaloa and Chihuahua, as a result of these and other complaints. A suspension had already been applied to its operations in Chihuahua because of complaints from the Baja Tarahumara.[53]

The firm's operations were "threatening to upset the public peace in the District of Cosalá," Fomento told the company in May 1891, when President Díaz suspended the survey in that district. Luis Martínez de Castro, apparently part owner of the company, replied that everyone in the mineral of Cosalá, where the complaints originated, had agreed to the survey, but because such places never lack "people of bad faith who, not being able to extract advantages from those having an interest in this survey, wanted to interfere in it in order to speculate with the delay in the issue of the title, and calling themselves injured, without any foundation whatsoever," they complained. Besides, he argued, it was unfair to suspend work in the whole district when no one was complaining elsewhere in it, since the company, "instead of appearing a threat as occurs elsewhere, everyone is pleased with the proceedings."[54]

There were those who disagreed. Agustín Zárate of Sinaloa's Badiraguato district was, in his words, the "victim of a scandalous deed" after he signed a composición contract with Sandoval and Pacheco in August 1891 in which he promised to pay the company US$100 in advance and US$300 a year later in exchange for the title to demasías he claimed. By 1893 the company had "been extinguished," Zárate told Fomento, and he still had not received his title. What was worse, the company had given his IOU for US$300 to a third party, who was now demanding that it be paid. Zárate was not the only victim of what amounted to a confidence game:

> Mr. Secretary, what just explanation could there be for this behavior? . . . Like myself, many other people who have made enormous

sacrifices to legalize the possession of their land with a title, quietly suffer similar outrages; but this state of affairs cannot continue as it is as long as we refuse to renounce our rights, [and refuse to] make ourselves complicit with those who, shielded by the law, have come to usurp from us our most honorable labor.[55]

When Fomento asked Sandoval for an explanation, he told them he would see that the land was paid for and the title delivered to Zárate and others who were buying the demasía. A title to the property in question was finally issued in October 1894, but because it was issued to the "possessors" it is not clear whether Zárate was among them.[56]

Fomento received a similar complaint from the holders of a rancho in 1894. In this case, the demasías found in the rancho of Santa Bárbara were to be divided into four lots and distributed to four different groups of owners. One of the four still had not received its titles three years after signing the contract with Sandoval and Pacheco—a situation, the individuals claimed, in which "a very large number of people find themselves."[57]

President Díaz's cancellation of the company's authorization to survey in response to the complaints was not welcomed everywhere, however. Representatives of the communities of Ronquillo, Palmar, La Palma, Carricitos, Gabriel, and Cahuinahuate in Badiraguato district tele-grammed the president in November 1891, within a few weeks of the cancellation, begging him to allow the surveys to continue there, "our settlements still pending. . . . Sr. Sandoval merits complete confidence." The president acceded, allowing the company to finish its work in those six places only to avoid damaging the interests of the property holders.[58]

By 1899 there were still eight properties on which composiciones arranged by Sandoval and Pacheco were pending; four were in Badiraguato district, two in Mocorito, and one each in the districts of Concordia and Sinaloa. All together, the company had surveyed 56,925 hectares and had identified 38,618 hectares of demasías. Because most of the work was done before the reforms mandated by the public land law of March 26, 1894, most of the composiciones "are very imperfect if they are judged, as they must be, in accord with the modern Land Law," a Fomento official concluded in 1899. "There scarcely appears, in general, the judicial description of the boundaries and a map, which is almost always incorrect." Sandoval told Fomento he did not want to perfect the pending composiciones because that would be burdensome; he preferred that the cases be handled as if they were regular surveys, titling him the one-third he was due. Fomento disliked this proposal, because the individuals who arranged the composiciones with the company had probably paid cash in advance, "without receiving any benefit because they have to arrange all over again the composiciones" of their land. It would not be good for the government, either, as the two-thirds it would get were already possessed (though not titled, of course), and "this causes problems in the future, as is well known." Nevertheless, since Sandoval was within his rights to drop the composiciones, Fomento sent form letters to the holders of the eight properties, giving them three months to arrange a composición directly with the ministry.[59]

The obvious objection to this decision was raised quickly by Cristino Sotelo, one of the property holders. Because Sandoval had already collected $200 (toward a total due of $500) for the demasías (1,393 hectares) the company found, he ought to be required to deliver the title as promised. Arranging a composición directly with Fomento would be too burdensome, Sotelo said. It would require a new survey, which would be a very costly undertaking because there were hundreds, if not thousands, of property owners surrounding the land in question. Fomento rejected Sotelo's appeal that he not be required to pay for a new survey and indicated no interest in helping him to recover his $200. The ministry received an identical appeal from another property holder who was also out $200, and reminded him that he could go to court to recover the money.[60]

Although Sandoval was never forced by Fomento to return any of the money he had received from the property holders, their land had at least been protected by Fomento from an obviously unscrupulous surveyor. When in 1902 Sandoval rather audaciously asked the ministry for land equivalent to one-third of what he had surveyed, he was turned down flat. By that time, too, Fomento had learned that he had taken money from all eight of the holders of the demasías, some of whom had by then managed to resurvey the land and arrange composiciones with the government.[61]

Telésforo García y Socios

When surveying began under one of Telésforo García's concessions in Sinaloa district in 1883, one property owner opposed the survey because it invaded her land. Her representative accompanied the surveyors as they worked under the legal supervision of the alcalde of San Pedro. Nothing further appears in the file, and there is no evidence that the company was titled any land as a result of the survey.[62]

Luis Martínez de Castro and the Sinaloa Land Company

The most active survey contractors in Sinaloa were Luis Martínez de Castro and the company to which he transferred his concession in 1905, the

Sinaloa Land Company. Although Martínez de Castro remained a share-
holder of the Sinaloa Land Company, the two contractors will be treated
separately in this section, partly to determine whether a U.S.–capitalized
survey company that even did business under its English name generated
different responses in the countryside than did Martínez de Castro when
he was operating independently and under his own name. Of the two
firms, the Sinaloa Land Company was the most active, surveying 101 lots
between 1905 and 1908. Of the 101 surveys, some kind of opposition
was recorded in 35, a proportion of 34 percent. A slightly lower proportion, 29 percent, was recorded in Martínez de Castro's surveys between
1899 and 1905. The latter surveyed 48 lots, with opposition appearing in
14. These figures and others are recapitulated in Table 13.[63]

Although the level of opposition that each surveyor provoked is about
the same in each case, a significant difference appears in certain variables
related to the oppositions. When landholders protested, Fomento was
more likely to reject the survey outright and deny compensation in the
case of the foreign firm. Twenty of the Sinaloa Land Company's 35 surveys in which opposition appeared were rejected by Fomento—the company never received compensation. The proportion of opposed surveys
thus rejected comes to 57 percent. Five of Martínez de Castro's opposed
surveys were rejected, a proportion of 35 percent. Moreover, opponents
were twice as likely to withdraw or abandon their opposition to Martínez
de Castro than to the foreign firm (4 out of 14, or 28 percent for the
former; 5 out of 35, or 17 percent for the latter.) Foreign ownership of
the Sinaloa Land Company may only have been a contributing factor to
its greater vulnerability, however. All of its surveys were undertaken after
Martínez de Castro's. Just as the Díaz government's control of surveyors
tightened considerably after 1900, it seems probable that its grip continued to stiffen in succeeding years, and that it became less and less willing
to risk popular discontent over the land issue.[64]

Data on opposition to surveys are given separately for each company
in the following paragraphs. Oppositions were grouped in three categories—those that resulted in Fomento's rejection of the survey and refusal
to compensate the company; those in which a court declared that the protesters had officially "desisted"; and all others. Since those in the last category are the ones in which evidence would most likely be found of
outright usurpation by the company with the connivance of the government, only they are described. See Table 13 for a comparative summary
of all categories.

The Sinaloa Land Company   Whenever the company offered to buy the
two-thirds of terreno nacional it had surveyed, anyone claiming posses-

sion of any of it had the right to buy it from the government before it could be sold to the company. Fomento's policy of offering such landholders the first right of refusal sometimes generated the initial protest. The company surveyed "Temporalidades" in Sinaloa district in 1905 and received title to its one-third in November 1906. A week later Fomento notified the possessors of the company's offer and asked them if they wanted to buy it at $1.30 per hectare in bonds. Four persons replied, saying they owned land jointly under an 1842 title and disputed the company's right to any property there, even though the file of the survey indicated that the same individuals not only failed to protest the survey but signed their names accepting it. The ministry, the landholders, and their lawyer (Jesús F. Uriarte, who was also representing the U.S. Sugar Refineries Company in its dispute with Navolato and Fomento) exchanged numerous letters arguing the validity of the title. Fomento stated that the document was no title at all but a mere "measurement" (i.e., part of a survey proceeding). The landholders requested and received several extensions to present other documents. This continued until August 1908, when Fomento turned down Uriarte's last request for another extension. The company's title to the two-thirds (442 hectares) was issued in September.[65]

Similarly, a representative of more than one hundred "joint owners and possessors" of San Joaquín in the same district claimed the two-thirds of terreno nacional under a title that was apparently never presented to the ministry. Fomento's offer to sell the land to the claimants remained open for nearly three years, until the ministry finally issued a title to the company.[66]

In 1907 two men representing the "Indians" of the Sinaloa district pueblo of Bacubrito asserted that Martínez de Castro, acting as the representative of the Sinaloa Land Company, took advantage of their ignorance to deprive them of their fundo legal. Two years ago, they said, the company came to survey their land, "making us believe that they did it in order to rectify the measurements of other land and acquiring some of our signatures. They came unexpectedly, and acting out of fear that if we did not sign something bad would happen to us, we signed where they said to, without of course the other Indians having any knowledge of what it was about."[67] Their title to the fundo legal was lost years ago, "since the Indians being naturally distrustful, the trustee of the title hid it so carefully that after his death it could not be found." Fomento pointed out that even though the pueblo was by law only entitled to 100 hectares for its fundo legal, that amount had been doubled in response to an earlier protest from the president of the ayuntamiento of Sinaloa district. Nevertheless, the ministry sent them a map produced by the company, asked them to mark the area covered by their title, and return it to Fomento after having it approved by the governor.[68]

Such flexibility by Fomento was not unusual. In one case, when a property holder objected to the company's survey of 1,878 hectares in Sinaloa district, title to all the surveyed land was issued to him, and the company took its compensation in land surveyed elsewhere.[69] In another, Pablo Cota telegrammed Fomento in August 1907 that he and the two hundred holders of a lot called "Baburia" only cultivated about 20 hectares each and needed the ministry's help "so as not to lose our home." They offered to buy their land at the published tariff for baldíos of $4 per hectare. Fomento agreed, and even offered to ask the company to take the one-third it was due in some other lot. The vecinos failed to reply to the offer, instead asking for extensions of the deadline for exercising their right of first refusal for the two-thirds. Two months after the company received title to its one-third in September 1909, Cota wrote again, this time on behalf of six hundred vecinos now holding just 10 hectares each. (Either he had forgotten his first story or the adult population had tripled in three years.) "How is it possible, Sr. Minister, that the bread and nourishment of our children be taken away from us?" He asked again to buy the two-thirds, but by then President Díaz had decreed the suspension of all terreno nacional sales.[70]

In two other surveys the company's title was issued after two bordering property holders who were given twenty-five days to take their protest to court failed to appear. In both cases more than three years elapsed between the date set for court appearances and the company's entitlement.[71] Four vecinos of Guasave objected to the company's rejection of their title to 1,515 hectares of land issued by the governor of Sinaloa in 1833; Fomento upheld the company's decision and advised them to take their complaint to court,[72] because the company's title was issued "without prejudice to a third party who has a better right."[73]

In the survey of "Santa María del Palmar" in Mazatlán district, the holders of the land reversed themselves twice, first approving the survey, then claiming it was never conducted, and later withdrawing their opposition. This was the largest single piece of land surveyed in the state of Sinaloa by any contractor. The Sinaloa Land Company measured 52,289 hectares, of which 9,496 turned out to be demasías. After Fomento offered the twenty-seven holders of the land the right to buy the two-thirds of the demasías, one of them replied that they were shocked by the offer because "there has not been any survey . . . nor has any authority nor any engineer called here that we know of." A 33,190-hectare lot that had once been demasía but was now protected by a title signed by President Díaz in 1891 was held by various people "who hold in common each lot independently of the others." He stressed that the land was not "community" even though each one was a "joint holder" among several holders, each being "owner of the lot that corresponds to them" but with no proprietary interest in any other lot. Thus, none of the vecinos care to buy "even a millimeter of those false demasías," which in any case were never surveyed. Six months later the same individual, along with twelve others, signed a letter stating that they did not object if the Sinaloa Land Company wanted to buy the two-thirds it surveyed. Fomento withdrew its sale offer to the company in 1911, however, because the firm failed to pay for the land within the time allotted.[74]

In this case the vecinos were probably confused at first about the location of the demasías, because the file indicated that three titles to a total of 42,434 hectares (including the vecinos' 33,190 hectares) were respected by the company, which took its one-third share (3,165 hectares) from among the true demasías of 9,496 hectares.

Luis Martínez de Castro   Of those surveys by Martínez de Castro in which some opposition appeared, all but six ended in either the rejection of the survey by Fomento or in a court determination that the complainant had dropped legal action.

Four of the six cases had in common the failure of the opponents to take their opposition to court, and consequently a title to Martínez de Castro (or the Sinaloa Land Company, which he occasionally designated as the recipient of titles for surveys conducted in his own name) in compensation for the survey was ultimately released. In two cases only one person objected; in one of them, two years elapsed between the survey and the issue of the title, while in the other, nine years elapsed before a title was signed. The remaining two surveys involved more opponents, but in one of them Martínez de Castro and Fomento agreed to let all the surveyed land remain "nacional," with the surveyor taking his compensation elsewhere. This was sometimes done in cases where opposition appeared, the purpose being to postpone the confrontation with irate landholders. The survey was approved, the land became "nacional" and thus went on the public land market, but resistance was presumbly deflated because the surveyor was out of the picture.[75]

Two other cases were considerably more complicated. In one the tenacious opponents were defeated after a five-year struggle that included an unsuccessful appeal to a Mexico City court. In the other, a single protester succeeded in convincing the company to let him buy all the surveyed baldío and demasías after six years of opposition.

In 1900 Fomento told the federal district judge that Martínez de Castro had failed to obtain the consent of the joint owners of a lot called "San

Javier" in Badiraguato district for the location of their titled land. More than a year later, twenty-four comuneros complained that when he had surveyed their land in 1899, Martínez de Castro had told them he would provide not only a title to their demasías but a "good discount on the present price." The complainants were reluctant to sign their approval because another comunero claimed that the new survey showed he owned a rancho that several of them inhabited. The comuneros offered to buy two thousand hectares for $1,000, intending to obtain a title "on behalf of two comuneros in order to then draw up the papers for a partnership and divide the land among our comuneros." Martínez de Castro replied with an offer to sell them four thousand hectares of demasías at fifty centavos each. After consulting they counter-offered with $2,000 for five thousand hectares, which Martínez de Castro rejected; he also threatened to sell all the land to an individual who would then charge "high rents." The comuneros later learned that Martínez de Castro had executed three deeds of sale to three men who were not comuneros, which they told Fomento they opposed.[76]

After reviewing the complaint, Fomento decided that the survey should be rejected and the comuneros invited to arrange a composición directly with the ministry, because

> in this case there has been an effort to exploit the ignorance and simplicity which the comuneros of San Javier reveal by means of their petition, and as possessors and longtime possessors and title holders, it seems that they should be protected and therefore the most appropriate action would be not to approve the survey. . . . [Martínez de Castro] either would still make them pay an excessive price or he would sell the land to others, . . . those with the most right to the land then have to pay rent or to abandon it.[77]

The comuneros took the case to court, arguing that the survey was inaccurate and arbitrary. The federal district judge ruled that the comuneros had failed to prove their charges and found in favor of the survey company. The plaintiffs' lawyer complained to Fomento of irregularities in both the survey and the litigation, but Fomento replied that it was beyond the ministry's power to change a "directed judicial verdict," which in any case was upheld by an appeals panel. Fomento said it had no choice but to approve the survey. (The survey could still be rejected if it was found to be defective, but this one was not, according to the ministry.) In September 1903, Martínez de Castro received his title to one-third of the land (2,475 hectares), as well as to another 1,501 hectares that he bought.[78]

In the second case Juan Castro Higuera, a vecino of the ranchería of El Limón, complained that Martínez de Castro, as he had elsewhere, ignored the rights of the "true owners of land" and violated legal procedure in the survey. Neither the real owners of the land nor those with property bordering it were ever summoned to the survey, which was conducted without the necessary judicial authorities. Castro Higuera added that the protesters' complaints were never recorded. Fomento, wondering why he had waited nine months to complain, replied that Castro Higuera was too late, because a title had already been issued to the surveyor for the one-third. Castro Higuera was not heard from again.

When the ministry sent a letter to the *poseedores* (holders) of the adjacent Aguacaliente de Zevada (within which the surveyor had located Castro's El Limón) asking them whether they wanted to exercise their right to buy the two-thirds of terreno nacional found there, Antonio Zevada offered to buy it. He had trouble proving he held the land, but after a meeting with the secretary of Fomento the latter informed Zevada that he did not have to prove possession if "that turns out to be very prolonged," but in that event he would have to pay a higher price—$6 per hectare. Two years later, in 1909, title was issued to the Sinaloa Land Company, which bought the property for $1.20 per hectare. Zevada complained, and to avoid litigation the company agreed to let him have the land. But Fomento could not approve the deal because of President Díaz's freeze on the sale of terreno nacional, and it recommended that the matter be settled in court.[79]

DURANGO
Rafael García Martínez

In July 1886 Col. Rafael García Martínez's company received the federal court's authorization to survey the municipalidad of Pueblo Nuevo, an area composed of five pueblos, a *congregación* (indian settlement), and twenty-eight ranchos.[80] By September the people of Pueblo Nuevo appeared to be on the verge of taking up arms in response to the survey, a development that García Martínez blamed on Juan Hernández y Marín, who had been the governor of Durango from 1871 until his ouster in 1877 by adherents of Gen. Porfirio Díaz. García Martínez's agent in Durango wrote to the colonel at his office in the Palacio Nacional that one of Hernández y Marín's men, Luis Silva, had arrived in Pueblo Nuevo on September 17, and "harangued the residents and incited them not to let their properties be snatched away by a Foreign Company and by the concessions and the indulgences of a clearly immoral Government. He went through the Town several times in a complete state of intoxication preaching the above-mentioned doctrines."[81]

Silva called a junta among the vecinos of several ranchos with the collaboration of the jefe político. On September 23 more than sixty men gathered in the "casa municipalidad." The jefe político, Luis Galindo, presided over the meeting while Silva urged the men to defend "the properties that they say must be theirs, properties being made use of by the Foreign Companies and the immoral Governments that have given them such concessions. He argued that the properties are the people's and not the Government's, which with titles or without are theirs to hold in spite of the Federation."[82]

Silva and Galindo collected money for a court battle, and Silva suggested that the land at stake be divided into small lots represented by shares. Whoever failed to pay the quota he set would be cut out of the group and would have to pay lots of money to the partnership, according to the obviously biased account of García Martínez's agent. García Martínez told Fomento that Silva only wanted to "exploit these poor people" and "instill revolutionary ideas in them." The agent then appealed to Gov. Juan Manuel Flores, who had led President Díaz's 1877 rebellion in Durango, for help.[83] Far from helping the survey company, however, Governor Flores told Fomento on October 18 that the company's claims were untrue. The survey of Pueblo Nuevo was proceeding "in perfect accord with the residents."[84] Nothing further was recorded about the residents' reaction to the survey of Pueblo Nuevo, suggesting that the opposition, perhaps deliberately incited by outsiders who had been deposed by the local Díaz forces in 1877, had either died away or been suppressed.

The residents were not always judicially informed of a survey, particularly if they were Indians and lived in settlements that had no recognized government. García Martínez's surveyors told the supervising judge in March 1887 that in eight pueblos in the partido of Mezquital the residents were "in their totality pure Indians," and because they also "lack authorities" no attempt was made to summon anyone to attend the proceedings; besides, none of the ranchos possessed a fundo legal. Nevertheless, Pedro Escobar appeared before the judge the following January to represent "the inhabitants and residents" of Xoconoxtle and Santa María de Ocotán, as well as the anti-Porfirista ex-governor of Durango, Juan Hernández y Marín, whom Escobar identified as the two pueblos' lawyer, who was in Mexico City arranging a settlement of their demasías. Escobar entered a formal protest but withdrew it nine days later, after the company agreed to respect the pueblos' borders. By May 1889, however, Hernández y Marín and the company had still not agreed on the status of land outside the pueblos' borders. García Martínez wanted Fomento to title him the company's full one-third anyway, but he was turned down because President Díaz had decided that the towns' claims were valid. The company got its title in July after 363,172 hectares described as land owned by "the Indian pueblos," to quote the title, were deducted from the area surveyed.[85]

The borders between states were sometimes as hard to identify and as subject to disputes involving the survey companies as were private property boundaries. During García Martínez's survey in 1892 of the municipalidad of Villa Ocampo in Partido Indé, the company placed an hacienda in Durango whose owners insisted was in Chihuahua. The governor of Chihuahua ordered the company's markers destroyed. Later, Fomento decided that the hacienda owners' complaint was in bad faith—they had never presented their titles for inspection, as requested, indicating that they either did not have titles or feared that upon exhibiting them "their total lack of right to the properties involved would be discovered." On these grounds the staff recommended rejecting their opposition, but the head of the ministry, Manuel Fernández Leal, decided to give the owners another month to present their titles. Eight months later—in February 1896—the owners still had not done so, and the Fomento staff recommended letting García Martínez receive title to one-third of the land under dispute, which had been held in reserve pending the presentation of the titles.[86]

Antonio Asúnsolo's Chihuahua survey company complained that García Martínez had invaded land it had surveyed in Chihuahua as well, prompting an investigation by Fomento that revealed both companies had surveyed "a great part of the same property." The government decided to settle the matter by letting the two companies solve the problem, which they did. In return for retracting his opposition to the survey and any claims that García Martínez had invaded Chihuahua, Asúnsolo received 2,000 hectares from García Martínez; the land was part of García Martínez's compensation for his survey of Villa Ocampo.[87]

After García Martínez's survey of Villa Ocampo, which began in 1892, the company received title to one-third of the baldío that was clearly not occupied, while some was held in reserve pending composición. Among the lands held in reserve were 6,742 hectares identified as "surplus" of the "ejidos de Villa Ocampo," the land left over after the company allotted the *villa* (town) a square league (1,756 hectares) for its ejido. In August 1896 Fomento wrote to the authorities of the villa, giving them two months to request of the ministry "arrangements to perfect your property," failing which the government would dispose of their land. By December there was still no reply and García Martínez was pressing Fomento for compensation for the land held in reserve, which his company had surveyed. The total amount due the company in compensation for various other lots in the municipalidad that were held in reserve pending composición came to 7,228 hectares. And because other property owners whose land had been held in reserve were now arranging composiciones, García Martínez wanted the 6,742 hectares in *sobrantes* (excess) to go toward his 7,228 hectares. By February 1897 Fomento was inclined to title the land to the company, because it was held by "muchas personas" who still had not bothered to respond to the ministry's offer, and "it is almost certain," an official wrote, "that they will never try to legalize their titles if they are not obligated to do so, remaining indefinitely in possession of those without the Government receiving their value."[88]

Fomento, continuing to ignore repeated requests from the company for title to the land, finally received its reply from the "vecinos of Villa Ocampo" in August 1902. To prevent its purchase by "some Company or individual who is a stranger to our Colony," the vecinos offered to buy part of the sobrante that had been held jointly as pasture. If the land were sold to anyone else, the vecinos would die of hunger, Villa Ocampo would be "reduced to Country Estate [Hacienda de Campo]" and all trade monopolized. Complaining that Fomento had been leasing the land to a Chihuahua rancher who was grazing livestock on it, they offered to take over the lease. By this time, former Durango governor Abel Pereyra had acquired the rights to the land still owed García Martínez. Pereyra had several meetings with Gov. Juan Santa Marina about how to "avoid problems" with the vecinos of Villa Ocampo, and as a result the governor suggested leaving a two thousand-meter strip of land between the sobrantes that Pereyra wanted to buy and the villa's fundo, and in addition to leave "free for the pueblo" other parts of the sobrantes, including some riverfront property. Pereyra's title to 2,616 hectares, less than half the sobrantes, was issued the following month.[89]

In response, some three hundred vecinos of Villa Ocampo sent a letter to President Díaz, informing him that they refused to believe the news that title to most of their land had been issued to Pereyra, "a very well-off person and the owner of a number of farms," and asking for relief. Fomento replied that Pereyra's title respected not only the ejido that had been designated by the original survey—to which the villa made no objection—but also an additional two-kilometer zone "out of equity toward you although it does not recognize your right to those lands." The petitioners were reminded that they were free to take the matter to court, as all titles were issued "without prejudice to a third party who may have a better right," according to the boilerplate language of most public land titles.[90]

Two years later Fomento informed Durango governor Estéban

Fernández that it was no longer "appropriate" for the vecinos of Villa Ocampo to continue holding in common land that was respected during García Martínez's survey as ejidos and fundo legal. Nor should they continue to enjoy those additional lands reserved for them outside the ejido without fulfilling the legal requirement that they be privatized under legislation permitting *cesiones gratuitas* (free transfers). Because the ministry was receiving requests from others for these lands, Fomento told the governor to order the subdivision of the ejidos and the fundo legal into lots, and to tell the vecinos to take the legal steps necessary to obtain private possession outside the ejidos. If no action was taken within three months, the land outside the ejidos would be sold by Fomento. The vecinos received a copy of the letter, and in December 1905, less than two months after it was written, they protested to Governor Fernández that the land at stake—ejidos, fundo legal, and the adjacent property—sustained "la vida del pueblo" and without it the town would die. They knew of requests by outsiders to acquire the land, but if a single person were to buy the land surrounding the villa the result would be "the death of the village and the ruin of the inhabitants." They asked that the land be transferred to individual vecinos instead, as permitted by law and suggested by Fomento, with each lot to be titled to whoever was holding it. Lots not currently held could be distributed by lottery or any other method the governor decreed. In February 1906 Fomento gave the vecinos a two-month extension to obtain the land under the cesiones gratuitas statute. For García Martínez's survey of the municipalidad, Fomento still owed the company 4,441 hectares. The titles were issued in 1910 and 1911, but whether the land was actually in the municipalidad is not clear, as compensation was sometimes made with land acquired by the government in other surveys, particularly when some of the originally surveyed land was claimed by third parties. The residents of Villa Ocampo, in any case, apparently delayed action on the disposition of the land in question until the eruption of the revolution, when such accounts were no doubt settled under the sponsorship of the insurgents.[91]

García Martínez's survey of the municipalidad of San Bernardo in the partido of El Oro caused protests by two hacienda owners, both of whom abandoned their opposition after accords were reached in which the company agreed to respect their boundaries.[92]

SONORA
Plutarco Ornelas

Sonoran landowner Manuel Elías complained in 1884 that the Ornelas company's survey of Arizpe district counted as baldío land to which he had claimed and acquired title. In June 1884, about eight months after the protest had been entered, the company rather cryptically told the federal district judge overseeing the survey that "because of difficulties that unfortunately befell the representatives of the Company, it was not possible to proceed as they should have with the respective operations." In his request for a new authorization to survey, the company's lawyer hinted to the judge that the "difficulties" were related to the universally contentious issue of public land claims pending at the time of a survey. The request was made with "the understanding that my clients solemnly affirm to respect both the properties of individuals and any others that legally may be acquired after the date of the concession."[93]

Nevertheless, the conflict with Elías over his public land claims continued. The company, in a considerably less humble petition to Fomento in December 1886—six months after it received title to the 20,284 hectares making up its one-third, as well as to the balance of 40,569 hectares, which it bought—asserted that Manuel Elías's acquisition of 7,500 hectares of public land (in the names of his children Eloisa, Carlota, and Francisco) had created a conflict. The reason was simple: All of the Elías's land was located on the same property that the company had just been titled. The company alone had a right to the land because its survey had preceded the issue of Elias's titles. The company also pointed out that Elías was not fulfilling his legal obligation to inhabit the land with at least one person for every 200 hectares, which Elías admitted.[94]

Fomento agreed with the company's reasoning and asked Elías to give back the titles and take 7,500 hectares elsewhere. Elías refused, contesting the company's right to the 60,853 hectares it had been titled. The company finally gave up and agreed to accept in compensation 7,500 hectares located outside the area it had surveyed. To link the 7,500 hectares with the surveyed area, the company bought 12,049 hectares of public land between the two places.[95]

The only other complaint recorded against Ornelas' survey was submitted by another public land claimant, but it too was of doubtful legitimacy. The protester had claimed the land after Ornelas's contract was issued, and got title to it after the company's title was signed. Fomento, not sympathetic, told the man to take the matter to court. His representative was Gen. Francisco Olivares, himself a survey contractor in Sonora.[96]

Mexican Land Colonization Company

In this company's survey of a 2,914-hectare parcel of baldío in Alamos district in 1892, the filing of two protests delayed the issue of titles for eight years. One protest was dismissed in 1898 after a trial, and when the company asked for its title Fomento noticed that another party had also protested the survey. The ministry first satisfied itself that the opposition had indeed been abandoned before issuing the title.[97]

CHIAPAS
Mexican Land Colonization Company

The survey of the Chiapas department of Motozintla by the Mexican Land Colonization Company was suspended in 1898 on orders of the jefe político of Motozintla, Salvador Palacios, because he feared the survey would diminish the ejidos of the pueblos of Mazapa and Motozintla. The company's frustrated local manager, O. H. Harrison, complained that the survey of the ejidos was "causing me considerable headaches" owing to the pueblos' "very defective" titles, which made it very hard to establish their boundaries. Usually, he said, this kind of thing was negotiated, but in these two cases "the people of these villages do not want to listen to reason." Even though the two towns' titles limited them to about 860 hectares each, they were occupying 12,900 hectares each, claiming all the land as ejido. Harrison's final offer to Palacios was to respect not only the quantity of land specified in the titles but also an additional 989 hectares for each pueblo, taking into account "present possession." The company further offered to sell to its holder any land outside that area for fifty pesos per *caballería* (one caballería equaled 43 hectares).[98]

Land Company of Chiapas

In response to an appeal from the company, Fomento wrote that conflicts between landholders and the Land Company of Chiapas that had arisen by 1909 would have to be aired in the local court. One exception was made, however. The government had offered land to some Guatemalans on the condition that they become Mexican citizens, and it felt obliged to keep its promise. Following the establishment of the international border under the Guatemalan border treaty these former Guatemalans had decided to become Mexican citizens, but the land they occupied had already been transferred to the Land Company of Chiapas. The company agreed to accept Fomento's offer of compensation with land elsewhere in return for letting the former Guatemalans keep what they held in San José Monte Sinai and other "villages and hamlets."[99]

In 1909 Fomento listed all the complaints pending against the company. There were eight, and it seems doubtful that any were settled before the overthrow of the Díaz government.[100]

1. Two men complained in 1907 that the company was attempting to take land they held.
2. The ayuntamientos of three pueblos in the department of Motozintla—San Isidro Siltepec, San Pedro Remate, and San Antonio la Grandeza—requested land they possessed. A study of the matter was pending.
3. The vecinos of three rancherías in the department of Soconusco asked that the company refrain from taking their land; the company had been asked for an explanation.
4. The vecinos of Juárez in the department of Pichucalco had lodged a complaint against the company, which Fomento determined was unfounded; they protested again, and a decision was pending.
5. Vecinos of Santa Rita and El Respiro in Soconusco asked the company to respect the land they held.
6. Various vecinos of Escuintla in Soconusco complained.
7. One man claimed land in Mocozintla.
8. A public land claimant in Tuxtla wanted to continue his claim after Fomento ordered it be suspended.

Bulnes Hermanos

When the Bulnes Hermanos' surveyors approached the Tabasco border along the Usumacinta River in their survey of Chiapas in 1885, people who were citizens of Tabasco but residents of Chiapas threatened the surveyors, who fled. The company, in recounting the incident to Fomento, said the opponents of the survey were illegally exploiting the fine timber of the region and selling it in Guatemala. The following year President Díaz ordered that approval of the Chiapas survey be withheld, "because of the irregularities with which they are covered, contrary to the letter and spirit of the relevant laws."[101]

TABASCO
Policarpo Valenzuela

When Policarpo Valenzuela surveyed part of the Gulf Coast region of Tabasco in 1892 one person filed a court action against the survey, stating that his property was not identified on the surveyor's map. He withdrew his opposition when the company altered the map to indicate his property.[102]

In October 1913, twenty years after Valenzuela's survey of the municipalidad of Macuspana, two men there wrote to Fomento in the name of eighty other *covecinos* (coresidents) asking that the land of "La Cochinera," which they all shared, be divided among them and converted to "legally titled property." Not only was the land wrongly included in Valenzuela's survey as baldío, they said, but it also was "measured imaginarily since no notice of a survey was given, which because it seriously affects our rights obliges us to make a legal protest." This is a curious complaint, not only because it occurred twenty years after the fact, but also because it indicates that the land continued to be occupied after the survey, suggesting that the occupants hardly suffered from it. Because titles were requested, the land clearly was not titled when surveyed, in which case Valenzuela had little choice but to call it baldío.[103]

Following approval of Valenzuela's survey of the municipalidad of Teapa in 1892 and the subsequent issue of title, the vecinos of Tecomajiaca, a barrio of the city of Teapa, went to court and claimed that the company had wrongly included some of the barrio's ejido as baldío. The case was settled out of court when Valenzuela agreed to return all the land claimed as ejido; he then exchanged the title he had already been issued for another that respected Tecomajiaca's ejido. This generous act did not meet the approval of the revolutionary government of Francisco Madero, because after the vecinos distributed the ejido land in sixteen-hectare plots among themselves, they never bothered to obtain titles for the land. Moreover, most of the so-called ejido had never been held by anyone. Fomento noted, however, that on the land that was possessed the occupants "pay their taxes religiously."[104]

Notes

PREFACE

1. Because two of the complaints against survey companies mentioned in chapter 4 were found in a classification other than deslindes, it is reasonable to infer that others were also submitted to Fomento and subsequently not filed under deslindes and not mentioned in the index summaries to other classifications, thus placing them beyond reach of the researcher confined to the deslindes section. The author wishes to acknowledge with thanks Professor Friedrich Katz's contribution of the copies of documents related to the two complaints not found in deslindes.

2. See the discussion in Wistano Luis Orozco, *Legislación y jurisprudencia sobre terrenos baldíos* (Mexico: Imp. de El Tiempo, 1895), pp. 342–47.

3. Ibid., p. 756.

4. See Daniel Cosío Villegas, gen. ed., *Historia moderna de México*, 9 vols. (Mexico: Editorial Hermes, 1955–72), *El Porfiriato: La vida económica,* "Moneda y bancos," by Fernando Rosenzweig, vol. 7, p. 866.

CHAPTER 1: LAND AND THE STATE IN PREREVOLUTIONARY MEXICO

1. Among the most useful interpretations of Mexico's economic growth and the state's activist role in this period are Fernando Rosenzweig, "El proceso político y el desarrollo económico de México," *Trimestre económico* 29 (1962): 513–530, and "El desarrollo económico de México de 1877 a 1911," *Trimestre económico* 32 (July/September 1965): 405–54; Cosío Villegas, *La vida económica;* Arnaldo Córdova, *La formación del poder político en México* (Mexico: Serie Popular Era, 1972); Carlos San Juan Victoria and Salvador Velázquez Ramírez, "El estado y las políticas económicas en el Porfiriato," in *México en el siglo XIX (1821–1910): Historia económica y de la estructura social,* ed. Ciro Cardoso (Mexico: Editorial Nueva Imagen, 1980); Sergio de la Peña, *La formación del capitalismo en México* (Mexico: Siglo Veintiuno Editores, 1975); François-Xavier Guerra, *México: Del antiguo régimen a la revolución,* 2 vols. (Mexico: Fondo de Cultura Económica, 1988), trans. of *Le Mexique. De l'ancien régime à la révolution* (Paris: L'Harmattan, 1985).

2. Juan Felipe Leal, "El estado y el bloque en el poder en México: 1867–1914," *Historia mexicana* 23 (April–June 1974): 703.

3. San Juan Victoria and Velázquez Ramírez, "El estado," pp. 280–85; De la Peña, *La formación*, p. 178.

4. Leal, "El estado," p. 715.

5. Leal, "El estado," p. 706; San Juan Victoria and Velázquez Ramírez, "El estado," p. 295.

6. John H. Coatsworth, "Del atraso al subdesarrollo: La economía mexicana de 1800 al 1910" (Chicago, 1984, unpublished manuscript), pp. 16–17, 364–65. Coatsworth estimates that over the course of the Díaz era, the government's contribution to GDP declined to about 7 percent. In absolute terms, government spending (measured in constant pesos) more than doubled. The drop in relative share is consistent with the state's objective of establishing the conditions for Mexico's economic growth. Porfirio Díaz and his científicos were not socialists; their goal was to boost *private* enterprise, and to the extent that they were able to achieve this objective while diminishing the state's overall participation (as measured by spending alone) is evidence of their success, not their failure.

7. San Juan Victoria and Velázquez Ramírez, "El estado," pp. 290–92.

8. Friedrich Katz, *The Secret War in Mexico* (Chicago: University of Chicago Press, 1981), pp. 5, 21–22.

9. Rosenzweig, "El desarrollo económico," p. 405; De la Peña, *La formación*, pp. 183–84.

10. Matias Romero, *Mexico and the United States* (New York: G. P. Putnam's Sons, 1898), pp. 7–9. The percentage was derived by reference to Mexico's total area of 198,720,100 hectares as given in Mexico, Secretaría de Fomento, Colonización e Industria, *Boletín de la Direccion General de Estadística*, no. 1 (Mexico: Imp. y Fototipia de la Sec. de Fomento, 1912), p. 75. Adding the two figures gives a total area before the treaties of 439,742,900 hectares, including Texas. Slightly less territory—195,820,100 hectares—is claimed in Mexico, Secretario de Programación y Presupuesto, Instituto Nacional de Estadística Geografía e Informática, *Agenda estadística 1984*, p. 18.

11. The decree was issued in December 1855, only three months after the resignation of President Antonio López de Santa Anna, who had promulgated similar legislation himself in 1853 and 1854. See the commentary in Orozco, *Legislación y jurisprudencia*, p. 291.

12. Francisco F. De la Maza, *Código de colonización y terrenos baldíos de la república mexicana* (Mexico: Ofc. Tip. de la Sec. de Fomento, 1893), pp. 658–59.

13. Mexico, Secretaría de Fomento, *Memoria de la Secretaría de Estado y del Despacho de Fomento, Colonización, Industria y Comercio de la República Mexicana escrita por el ministro del ramo, C. Manuel Siliceo, para dar cuenta con ella al Congreso Constitucional* (Mexico: Imp. de Vicente García Torres, 1857), p. 39.

14. Mexico, Secretaría de Fomento, *Memoria*, 1857, pp. 39–40.

15. De la Maza, *Código*, pp. 623–24; 641–45.

16. Mexico, Secretaría de Fomento, *Memoria*, 1857, p. 48.

17. Ibid.

18. De la Maza, *Código,* pp. 729–35.

19. Unless otherwise specified, global figures on the transfer of public land were taken from Mexico, Secretaría de Fomento, *Memoria presentada al congreso de la unión* . . . 1877–82, 1883–85, 1892–96, 1897–1900, 1905–7, 1907–8, 1908–9 (Mexico: Sec. de Fomento); and Secretaría de Fomento, Dirección General de Estadística, *Anuario estadístico de la república mexicana* 1893, 1897, 1901, 1906 (Mexico: Sec. de Fomento).

20. Don M. Coerver, "The Perils of Progress: The Mexican Department of Fomento during the Boom Years, 1880–1884," *Inter-American Economic Affairs* 31 (Autumn 1977): passim, 44.

21. ATN. CHIH 77415, pp. 57–58.

22. De la Maza, *Código,* p. 962.

23. ATN. DUR 9053, cuaderno 1, Fomento to Justicia, Mar. 24, 1887. Increased government revenue would be realized in various ways: the sale of its two-thirds share, the pressure exerted by the companies on squatters who would file revenue-generating land claims, and the economic growth generated by a bullish land market.

24. ATN. CHIAP 84797, pp. 76–77.

25. Mexico, Secretaría de Fomento, Colonización, Industria y Comercio, *Memoria presentada al congreso de la unión* . . . *corresponde a los años trascurridos de diciembre de 1877 a diciembre de 1882* (Mexico: Ofc. Tip. de la Sec. de Fomento, 1885), 1:39–40.

26. Ibid.

27. Mexico, Secretaría de Fomento, *Memoria,* 1877–82, 1:40.

28. Adolfo Díaz Rugama, *Prontuario de leyes, reglamentos, circulares y demás disposiciones vigentes relativas a* . . . *la Secretaría de Fomento* . . . (Mexico: Eduardo Dublan, 1895), p. vii. The uncertainty of boundaries continued to plague Mexican property relations for many decades. As recently as the 1960s, the Mexican government's announcement that it would undertake a modern cadastral survey was welcomed on the grounds that it would eliminate one of the "causas fundamentales de las interminables controversias que agobian el agro mexicano." Manuel Aguilera Gómez, *La reforma agraria en el desarrollo económico de México* (Mexico: Instituto Mexicano de Investigaciones Económicas, 1969), p. 153.

29. ATN. CHIH 1.71/10, court file.

30. ATN. CHIH. 1.322/1, Charles to Fomento, Sept. 20, 1879.

31. Before 1894, individuals had two means of acquiring public land: by the claims method or by arranging a composición with the government. After the passage of the land law of March 26, 1894, public land could also be purchased outright from the government without going through the cumbersome claims process. Such sales were limited to what was specified in the act as "terreno nacional," which by definition was surveyed public land. Composiciones were exclusively the privilege of individuals who already possessed the land in question. See Orozco, *Legislación y jurisprudencia,* pp. 403–38.

32. CPD. 41/7.15/245–46, 263–64.

33. ATN. DUR 1.71/20, pp. 39–48.

34. ATN. DUR 1.71/30, Casasús to Fomento, June 14, 1888.

35. José Covarrubias, *Varios informes sobre tierras y colonización* (Mexico: Imp. y Foto. de la Sec. de Fomento, 1912), pp. 10–15, 6.

36. Mexico, Secretaría de Fomento, *Informe que rinde el Secretario de Fomento a la honorable Cámara de Diputados sobre colonización y terrenos baldíos.* (Mexico: Ofc. Tip. de la Sec. de Fomento, 1885), pp. 155, 191.

37. ATN. CHIAP 1.71/15, pp. 3–24.

38. E. Kozhevar, *Informe sobre la república mexicana presentada al Consejo de Tenedores de Bonos Extranjeros*, trans. Joaquín D. Casasús (Mexico: Ofc. Tip. de la Sec. de Fomento, 1887), pp. 44–45. The last sentence quoted is a reminder that large landowners, as well as peasants, had ample reason to fear the survey companies. Not only would an accurate survey of their holdings threaten to shrink their domain, but it might also provide an incentive for foreign investors to bid up the price of choice land.

39. Kozhevar, *Informe*, p. 54. Government outlays had vastly exceeded income beginning in 1881, and by June 30, 1884, the deficit over the past three years had mounted to $19 million—about 40 percent of all federal government expenditures in 1882–83. See Kozhevar, pp. 58–59.

40. In the mid-1890s, the Mexican government was selling its surveyed public lands for an average of $1.02 per hectare, about 2.6 times what it charged for unsurveyed public lands, which were going for an average of $0.39 per hectare. See Mexico, Ministerio de Fomento, *Anuario estadístico de la república mexicana, 1897* (Mexico: Ofc. Tip. de la Sec. de Fomento, 1898), pp. 371, 376.

41. Mexico, Secretaría de Fomento, *Memoria*, 1877–82. 1:40.

42. De la Maza, *Código*, pp. 992–93.

43. CPD. 15/4/1770 (Preciado to Díaz); 41/7.16/477–78 (Díaz to Preciado).

44. The figure of 21.2 million hectares actually covers the period 1883–1907. The government reported that no compensations were issued to survey companies from 1878–82, nor for the year 1908. The latter year was the latest for which land transfer data were published. The area of Mexico was given as 198,720,100 hectares in the source cited in n. 10.

45. Fifty is an estimate based on the fact that in the six states represented in this study, twenty-six companies received a total of 10.5 million hectares, or almost exactly half of all the land given in compensation throughout the country during the Porfiriato. Because half the land was given to twenty-six companies, it seems unlikely that more than fifty companies ever received compensation. Also, some of the companies in the six study states were active in other states, so the number is probably well under fifty.

46. Miguel Mejía Fernández's assertion that communal land was usurped "en gran escala . . . en virtud de la acción vandálica de las compañías deslindadoras," is typical. See his *Política agraria en México en el siglo XIX* (Mexico: Siglo Veintiuno Editores, 1979), p. 231. Also see Helen Phipps, *Some Aspects of the Agrarian Question in Mexico: A Historical Study* (Austin: University of Texas Bulletin No. 2515, Apr. 15, 1925), pp. 111–15; George McC. McBride, *Land Systems of Mexico* (New York: American Geographical Society, 1923), pp. 74–81; Andrés Molina Enríquez, *Los grandes problemas nacionales* (Mexico: Imp. de A. Carranza e

Hijos, 1909; Mexico: Centro de Estudios Históricos del Agrarismo en Mexico, 1984), pp. 87, 89; Daniel Cosío Villegas, gen. ed., *Historia moderna de México*, 9 vols. (Mexico: Editorial Hermes, 1955–72), *El Porfiriato: La vida social,* by Moisés González Navarro, vol. 4, p. 188; Jesús Silva Herzog, *El agrarismo mexicano y la reforma agraria* (Mexico: Fondo de Cultura Económica, 1959), p. 117; Nathan L. Whetten, *Rural Mexico* (Chicago: University of Chicago Press, 1948), p. 87; Eyler N. Simpson, *The Ejido: Mexico's Way Out* (Chapel Hill: University of North Carolina Press, 1937); Marco Bellingeri and Isabel Gil Sánchez, "Las estructuras agrarias bajo el Porfiriato," in *México en el siglo XIX (1821–1910): Historia económica y de la estructura social,* ed. Ciro Cardoso (Mexico: Editorial Nueva Imagen, 1980), p. 315. In the first 170 pages of his two-volume history of the revolution, Alan Knight reviews the political, social, and economic conditions of the Porfiriato. The survey companies are accorded exactly three sentences; typically, his figures on public land transfers come not from the documents published by the government but from the partisan and semi-journalistic accounts of McBride and F. González Roa. See his *The Mexican Revolution,* vol. 1, *Porfirians, Liberals and Peasants* (Cambridge: Cambridge University Press, 1986), p. 95.

47. Enrique Florescano, "Ensayo de Interpretación," in *Latin America: A Guide to Economic History, 1830–1930,* ed. Roberto Cortes Conde and Stanley Stein (Berkeley: University of California Press, 1977), p. 440.

48. François Chevalier, *L'Amerique Latine de l'indépendance a nos jours* (Paris: Presses Universitaires de France, 1977), p. 291.

CHAPTER 2: FOSTERING DEVELOPMENT

1. David W. Walker, *Kinship, Business and Politics: The Martínez del Río Family in Mexico, 1824–1867* (Austin: University of Texas Press, 1986), pp. 22–26, 217–20.

2. Roderic A. Camp, *Entrepreneurs and Politics in Twentieth-Century Mexico* (New York: Oxford University Press, 1989), pp. 11–12, 33, 250–51.

3. Ibid., p. 15.

4. Ibid., pp. 78–79.

5. Alan Knight describes the expansion of middle class in *The Mexican Revolution* 1:42–44.

6. Guerra, *México* 2:336.

7. An authorization was little more than a letter permitting an individual or firm to conduct a survey. A contract was more formal and appeared to require the signature of the surveyor. In this study, contract and authorization are used interchangeably.

8. Orozco, *Legislación y jurisprudencia,* p. 804.

9. Ibid., p. 807.

10. Ibid., pp. 814–15. If a surveyor ever requested or received compensation in the form of the land's "value" instead of in land itself, no evidence of such a transaction turned up in the archival record.

11. Rejections totaled four in Durango, three in Sinaloa, and one each in Chiapas and Tabasco. No reason was given in three cases; in the others,

the petitioners were told that enough contracts had already been let for the areas requested.

12. ATN. SIN 84793, pp. 107, 111.

13. ATN. CHIAP 1.71/22, pp. 1–5.

14. Orozco, *Legislación y jurisprudencia,* p. 821.

15. The reasons for the failure of colonization have been enumerated at length elsewhere. See, among others, Diego G. López Rosado, *Historia y pensamiento económico de México,* 2 vols. (Mexico: UNAM, Instituto de Investigaciones Económicos, 1968), 1:202–3, and Moisés González Navarro, *La colonización en México, 1877–1910* (Mexico: Talleres de Imp. de Estampillas y Valores, 1960).

16. Mexico, Secretaría de Fomento, *Memoria,* 1897–1900, p. 7; Guillermo Wodon de Sorinne, *La colonización de México* (Mexico: Sec. de Fomento, 1902), p. 14. Also see Mexico, Sec. de Fomento, *Memoria,* 1907–8, p. xxiv; and Covarrubias, *Varios informes,* pp. 28–31, 347–48.

17. De la Maza, *Código,* p. 932.

18. ATN. DUR 113062, pp. 75–79; 1.71/37, memo, Sept. 7, 1893.

19. On this point see ATN. DUR 1.71/37, memo, Sept. 7, 1893, and TAB 1.71/37, p. 127. Copies of all important Fomento correspondence, however, were routinely shared with the state governor and the district judge.

20. ATN. CHIH 1.71/119, Juez del Distrito de Chihuahua, Jan. 11, 1883.

21. ATN. SIN 1.71/43, court file.

22. ATN. CHIH 1.71/119, court file.

23. ATN. SIN 1.71/47, court file.

24. Since Fomento was hiring the companies precisely to determine the quality as well as the location of the public lands, the ministry often found it nearly impossible to adequately assess whether the government would wind up with two-thirds of a tract whose only valuable portion would be given to the survey company. When a Fomento official approved a division proposed by Policarpo Valenzuela in Tabasco, he noted that what was reserved for the government appeared to be of equal quality to the company's share, "although it would be difficult to be sure because the maps are not highly detailed." ATN. TAB 1.71/24, Memo, Apr. 8, 1891. Companies nearly always specified the land they wanted, but in the Mexican Land Colonization Company's survey of Chiapas, the company let Fomento choose its two-thirds share, even though the government "always picks the best," its lawyers complained. When a Fomento agent was sent to Chiapas to take possession of land in Pichucalco in the government's name, however, he wired back that it was the worst in the whole department. His superiors replied that the government had already chosen the property and he was expected merely to verify its limits. See ATN. CHIAP 1.71/34, Velasco to Fomento, Jan. 8, 1902; Ismael Loya to Fomento, May 26, 1902; Fomento to Loya, May 29, 1902. When Policarpo Valenzuela failed to indicate the land he wanted to keep in Tabasco, the government picked it for him, leaving the best for itself by order of the secretary. ATN. TAB 1.71/11, pp. 38–39, TAB 1.71/6, p. 12.

25. ATN. CHIH 1.71/56, pp. 140–84, 189, 207–19.

26. ATN. CHIH 75556, pp. 397, 399, 401–7.

27. ATN. CHIH 54653, cuaderno 1, Valenzuela to Fomento, Jan. 8, 1885.

28. ATN. SON 1.71/32, pp. 78–113.

29. ATN. TAB 1.71/11, pp. 49–50.

30. ATN. DUR 113058, court file. Representatives of two of these pueblos, Jaconoxtle and Sta. María de Ocotán, appeared before the federal judge in Durango to oppose the survey, then withdrew their opposition after the company agreed to respect certain land for which the towns held titles, pending their inspection and approval by Fomento.

31. ATN. SIN 1.71/40, memo, Nov. 20, 1891, and title, Nov. 24, 1891.

32. No opposition to the survey was recorded, though Fomento was satisfied that all titled properties and those pending entitlement were respected. ATN. TAB 1.71/1, pp. 55–57.

33. ATN. TAB 1.71/4, p. 18.

34. Not counted is Bulnes Hermanos, whose titles were revoked by the Díaz government. See chapter 3.

35. Huller was identified in a legal document in 1885 as "of German origin, a vecino of the State of Sonora." NOT. José María Velásquez (no. 732), July 29, 1885, pp. 153–54. But Cosío Villegas, *Historia moderna* 2:1183 calls him "alemán naturalizado norteamericano" ("a naturalized North American German").

36. The six were Francisco Olivares, the Mexican Land Colonization Company, Manuel Peniche, the Sinaloa Land Company, Luis Huller and the International Company.

37. It is of course possible that the Mexicans in these cases were acting as *prestanombres* (name lenders) to initially shield the foreigner's participation. And, as already indicated, those companies for which no evidence of foreign participation appeared in the record may in fact have been held in whole or in part by foreigners using prestanombres.

38. Sinaloa was excluded because 156 titles were issued for that state, more than for all five other states combined (104). Most of Sinaloa was not surveyed until after 1900, long after surveying in the rest of the country had practically ended. The zones surveyed there, and thus the parcels titled, were much smaller than elsewhere, as noted earlier.

39. A few general guidelines are mentioned in an 1863 decree. See De la Maza, *Código*, pp. 737–39. In 1884 Fomento instructed the federal district judges to include, with the files of surveys forwarded to Fomento for its approval, a report by the engineers who conducted the surveys describing the system used to measure lines and angles, and how the area was calculated, along with all the results, to avoid having to request clarifications and changes. See De la Maza, *Código*, pp. 958–59.

40. ATN. SON 1.71/18, memo, May 3, 1894.

41. ATN. SON 1.322/11, Fomento to Agustín González, Apr. 12, 1888. Even though the International Company recognized that it did not have to do so, it did indicate the location of private properties on its maps of Sonora. See ATN. SON 1.71/4, pp. 81–85.

42. ATN. CHIAP 1.71/1, pp. 84–109.

43. Ibid. pp. 119–24, 127–28, 131–35; 1.71/24, pp. 33–35. For Fomento's

long-running disputes with this company over the accuracy of its Chiapas survey, see chapter 3.

44. ATN. TAB 1.71/24, Vicente Dardón to Fomento, Dec. 30, 1889; Feb. 3, 1890. As a result of the complaint, Valenzuela's survey was suspended and Dardón was hired to correct it; he failed to do so, however. See chapter 3.

45. ATN. CHIAP 1.71/1, José Covarrubias to Sr. Ministro, Jan. 12, 1895.

46. ATN. TAB 1.71/6, R. Mendoza Cuesta to Ing. Máximo Alcalá, May 7, 1912.

47. ATN. DUR 76139, pp. 99–101.

48. ATN. SON 78729, pp. 137–45.

49. ATN. TAB 1.71/11, pp. 38–39.

50. ATN. CHIH 1.71/10, court file.

51. ATN. SON 78729, court file, pp. 12–35; SON 1.71/4, pp. 56–79.

52. ATN. CHIAP 1.322/8, Juan A. Navarro to Fomento, Oct. 23, 1891.

53. CPD. 14/1/238–42.

54. ATN. CHIH 1.322/12, pp. 42–49.

55. ATN. TAB 1.71/18, court file.

56. ATN. TAB 1.71/32, pp. 33–35.

57. ATN. CHIH 1.71/75556, p. 397.

58. ATN. SIN 1.71/84, Francisco de P. Alvarez to Martínez de Castro, July 18, 1906.

59. Mark Wasserman, *Capitalists, Caciques, and Revolution: The Native Elite and Foreign Enterprise in Chihuahua, Mexico, 1854–1911* (Chapel Hill: University of North Carolina Press, 1984), pp. 73–75.

CHAPTER 3: STATE MANAGEMENT OF THE SURVEYS

1. The state's devotion to economic development was reflected in the early Díaz administration's decision to make the cutting of telegraph wires punishable by death. In Díaz's words, if the cutting "occurred on a plantation the proprietor who failed to prevent it should be hanged to the nearest telegraph pole. . . . We were harsh. . . . But it was all necessary then to the life and progress of the nation." James Creelman, "Díaz Speaks," in *The Age of Porfirio Díaz: Selected Readings,* ed. Carlos B. Gil (Albuquerque: University of New Mexico Press, 1977), pp. 80–81.

2. One of the most strikingly recurrent themes in the files of Porfirio Díaz's correspondence, as preserved in the Colección General Porfirio Díaz, was the government's almost obsessive interest in keeping peace in the countryside to avoid giving foreign investors grounds for diverting funds elsewhere.

3. Lorenzo Meyer, "Historical Roots of the Authoritarian State in Mexico," in *Authoritarianism in Mexico,* ed. José Luis Reyna and Richard S. Weinert (Philadelphia: Institute for the Study of Human Issues, 1977), pp. 4–5; Daniel Cosío Villegas, gen. ed., *Historia moderna de México,* 9 vols. (Mexico: Editorial Hermes, 1955–72), *El Porfiriato: La vida política interior,* pt. 2, p. 313; Don M. Coerver, *The Porfirian Interregnum: The Presidency of Manuel González of Mexico, 1880–1884* (Fort Worth: Texas Christian University Press, 1979), pp. 2, 298.

4. While this section contains some data on Fomento's efforts to protect the rights of property holders (especially indigenous peoples) who lacked titles, the manner in which it responded to complaints of usurpation by landholders is treated in chapter 4.

5. ATN. CHIH 54653, cuaderno 1, Fomento to José Valenzuela, Jan. 5, 1885.

6. ATN. CHIH 97514, Fomento to José Valenzuela, Aug. 24, 1885.

7. ATN. DUR 113060, memo, Jan. 30, 1895.

8. ATN. SIN 1.71/55, memo, Oct. 28, 1892; Fomento to Martínez del Río, Nov. 10, 1892; SIN 1.71/51, memo, Aug. 15, 1892.

9. That the appropriate variable was not the location of the survey but the timing can be supported with evidence that Fomento was similarly critical of surveys conducted in other states after 1900. A survey of part of the department of Chilón in Chiapas by Martínez de Castro was returned three times in 1902 before it was finally approved; it had omitted evidence of legal procedures to protect property holders. See ATN. CHIAP 78628, passim.

10. See, for example, ATN. SIN 1.71/134, p. 31; 1.71/148, p. 60; 78405, memo, Oct. 24, 1900.

11. ATN. DUR 9053, cuaderno 1, p. 56.

12. ATN. DUR 1.322/19, memo, Sept. 19, 1923.

13. ATN. DUR 1.71/6, pp. 15–18.

14. ATN. CHIH 1.71/112, Pacheco to Gómez del Campo, Nov. 17, 1887. Also see ATN. DUR 113060, Fomento to Rodolfo Reyes, Mar. 1, 1905, which informed García Martínez's representative that if the company invaded titled land, "it is evident that it has no right to any compensation."

15. ATN. DUR 76139, pp. 259–60; DUR 1.71/38, Tinoco to Fomento, Sept. 1, 1893, "Informe acerca de las operaciones de rectificación de medidas del Partido de Santiago Papasquiaro." In this case Tinoco was compensated with one-third of the land he surveyed.

16. ATN. SON 78729, pp. 75, 131.

17. Ibid., p. 42.

18. Ibid., pp. 137–45.

19. Ibid., p. 75; title, Oct. 27, 1887. There is no record of how Peniche compensated the government for the resurvey, if he ever did so.

20. ATN. TAB 1.71/28, pp. 256–62.

21. ATN. TAB 1.71/14, pp. 14–18; TAB 1.71/28, pp. 280–81.

22. ATN. TAB 1.71/28, pp. 250, 256–62.

23. ATN. SON 1.71/37, pp. 40–66, 68.

24. Alfonso Luis Velasco, *Geografía y estadística de la república mexicana,* vol. 20, *Geografía y estadística del estado de Chiapas* (Mexico: Ofc. Tip. de la Sec. de Fomento, 1898), pp. 29–30.

25. ATN. CHIAP 1.71/25, pp. 62–63.

26. ATN. CHIAP 1.71/1, pp. 308–13.

27. See the discussion of survey methods in chapter 2.

28. ATN. CHIAP 1.71/1, pp. 84–109.

29. Ibid., p. 125.

30. Ibid., pp. 127–28.

31. Ibid., pp. 131–35.

32. Ibid., pp. 136–42.

33. Ibid., pp. 147–48, 164, 193–94, 215.

34. Ibid., pp. 267–75, 308–13, 353–56.

35. When the company was titled that land in 1890, Fomento itself had set the value of the highest grade of unclaimed public land in Chiapas at $1.65 per hectare; if all the company's land there was first class, it would have been worth $57,750. To Covarrubias, then, a survey added as much as 73 percent to the value of baldío, a topic treated in chapter 5.

36. ATN. CHIAP 84797 Anexo.

37. This amounted to one-sixth of all the land surveyed, because under its contract the company received one-third in compensation, and was authorized to buy three-fourths of the remainder at $1.10 per hectare.

38. ATN. CHIAP 1.71/24, pp. 130–31, 150–53, 176–77, 244–48.

39. ATN. CHIAP 1.71/34, Velasco to Fomento, Jan. 8, 1902.

40. ATN. CHIAP 1.71/34, Loya to Fomento, May 26, 1902; Fomento to Loya, May 29, 1902; Fomento to Velasco, Dec. 29, 1902; Velasco to Fomento, Jan. 7, 1903; memo, Feb. 28, 1903; Fomento to Velasco, Mar. 6, 1903; Plácido Gómez to Fomento, Mar. 25, 1903.

41. ATN. CHIAP 1.71/5, pp. 2–44; CHIAP 84796, pp. 295–96; CHIAP 1179, pp. 1–2.

42. ATN. CHIH 97449, pp. 58, 97–98, 144.

43. Cases involving conflicts with local officials over protests by landholders will not be repeated here, as they are noted in chapter 4.

44. The period still lacks an overall treatment of the operation of the Mexican government at the federal and regional levels. For the view of the government as a dictatorship in which power was highly concentrated in the executive, see Molina Enríquez, *Los grandes problemas nacionales,* pp. 87–88, who notes that Díaz controlled the selection of all federal judges and even intervened in their decisions in "casos especiales." Justo Sierra, the regime's científico apologist, observed that it was necessary for Díaz to "direct actively the political bodies of the country, including state legislature and governors." See his *The Political Evolution of the Mexican People,* trans. Charles Ramsdell (Austin: University of Texas Press, 1969), p. 365. According to Carleton Beals, Díaz by 1892 had perfected a "domination as absolute as the rule of the old-style Turkish Sultan." *Porfirio Díaz: Dictator of Mexico* (Philadelphia: J. B. Lippincott Co., 1932), p. 297. More recent scholarship stresses that Díaz did not effectively consolidate his power until after 1888; see chapter 3.

45. For a discussion of this practice as it related to the overall strategy of the survey company policy, see chapter 4.

46. ATN. CHIH 97374, pp. 21, 24–25, 31.

47. ATN. CHIH 54653, cuaderno 2, pp. 27–28; CHIH 75669, pp. 1–4.

48. ATN. DUR 9053, cuaderno 1, García Martínez to Fomento, Dec. 18, 1890; Juez de Distrito to Fomento, Jan. 8, 1891; Fernández to Fomento, Nov. 17, 1893; Fomento to Juez, Dec. 2, 1893.

49. ATN. CHIAP 1.71/24, p. 53.

50. ATN. TAB 1.71/32, p. 29–36.

51. ATN. TAB 1.71/1, p. 3; TAB 1.322/2, p. 215.

52. ATN. TAB 1.71/1, pp. 45–46, 49–52.

53. ATN. TAB 1.322/2, p. 218.

54. ATN. DUR [sic] 1.71/6, pp. 29, 32. CHIH 1.322/12, p. 11.

55. ATN. DUR 1.322/18, García to Fomento, July 8, 1883; Fomento to Juez, July 17, 1883.

56. ATN. DUR 1.71/6, p. 1. See Appendix A for details on the ownership of the new company.

57. ATN. DUR 76142, memo, July 15, 1890; Fomento to Calderón, July 29, 1890; Calderón to Fomento, July 30, 1890.

58. ATN. DUR 113060, Valenzuela and Fernández to Fomento, Nov. 3, 1894; DUR 113070, pp. 101–3, 108–10; DUR 1.71/20, pp. 21–23.

59. ATN. SON 78729, pp. 77, 80, 119, 43–58, 149, 201–2.

60. See the discussion of this case in the first section of this chapter.

61. ATN. SON 1.71/31, contract between Conant and Mexican Land Colonization Company, Sept. 28, 1891.

62. ATN. SIN 84792, pp. 20–23.

63. ATN SON 1.322/11, pp. 50–51, 53–55.

64. ATN. TAB 1.71/24, passim.

65. ATN. SON 78729, pp. 165–82.

66. ATN. SON 1.71/38, memo, Feb. 25, 1891; Fomento to Quijano, June 29, 1891.

CHAPTER 4: PROPERTY RIGHTS IN A MODERNIZING ECONOMY

1. ATN. SON 1.71/38, contract, Nov. 26, 1890.

2. CPD. 11/13/6084–85.

3. ATN. SON 1.71/31, Emilio Velasco to Fomento, May 6, 1889; SON 1.71/21, p. 22.

4. De la Maza, *Código,* pp. 932–33.

5. Ibid., pp. 992–93, 996.

6. ATN. DUR 9053, cuaderno 1, "Informe del apeo de la 1a zona de Durango"; DUR 1.322/17, M. Romero Rubio to Fomento, May 11, 1883, and June 11, 1884.

7. ATN. CHIAP 1.71/24, pp. 156–63.

8. De la Maza, *Código,* pp. 945–46, 959–60, 998, 1064–65.

9. ATN. SON 1.71/26, Fomento to Juez de Distrito, June 8, 1887.

10. ATN. TAB 1.71/4, memo, Oct. 10, 1891.

11. ATN. CHIAP 1.71/1, pp. 223–25, 211.

12. ATN. CHIAP 1.71/9, Velasco to Fomento, Mar. 9, 1891; Fomento to Juez, June 26, 1891; Velasco to Fomento, July 7, 1891; CHIAP 1.71/1, pp. 223–25.

13. CPD. 14/7/3158.

14. ATN. DUR 1.71/30, Casasús to Fomento, June 14, 1888.

15. ATN. DUR 1.71/30, Fomento to Casasús, July 4, 1888. For details on the process of composición, see the earlier section on Sinaloa.

16. Only for post-1900 Sinaloa was an effort made to compare total surveys to surveys in which opposition was expressed. Table 13 indicates that 49 out of 149 surveys, or 1 out of every 3, were opposed by a least one person. This is a surprisingly high rate in a period when the regime was becoming more authoritarian.

17. The vecinos of Huejotitán in Balleza, Chihuahua, complained of armed assaults by a surveyor hired by a group that bought land from one Ramón Guerrero, a survey company owner.

18. ATN. CHIH 75669, court file.

19. ATN. SIN 1.71/47, court file.

20. ATN. TAB 1.322/2, pp. 150–53.

21. Ibid., pp. 148–49, 153.

22. CPD. 10/3/1394; 10/4/1719.

23. ATN. CHIAP 1.322/2, pp. 130–33.

24. ATN. TAB 1.322/2, pp. 130–33, 145, 147.

25. CPD. 41/3.5/298, 438.

26. CPD. 11/7/3458, 3484.

27. CPD. 41/6.13/222–23.

28. ATN. TAB 1.71/37, pp. 10, 15–20.

29. Ibid., pp. 37–42, 71, 77, 80, 104–5.

30. ATN. TAB 1.71/26, pp. 1–6, 13. 1.71/16, 424–28.

31. Ibid., pp. 1–4, 7–8.

32. Ibid., pp. 9–13, 48–49.

33. John R. Southworth, *El estado de Sinaloa, Mexico* (San Francisco: Press of the Hicks-Judd Co., 1898), pp. 43–47.

34. ATN. SIN 1.71/26, pp. 58–61, 67.

35. Ibid., pp. 71–75.

36. Ibid., pp. 201–50.

37. ATN. SIN 1.71/63, pp. 3–13.

38. ATN. SIN 1.71/26, pp. 393–97.

39. The price of public land subject to denouncement in Sinaloa leaped 233 percent from 1905 to 1907; because these prices were set by the federal government, they undoubtedly lagged behind market prices. For public land prices, see Aniceto Villamar, *Las leyes federales vigentes sobre tierras, bosques, aguas, ejidos, colonización y el gran registro de la propiedad* (Mexico: Herrero Hermanos, 1910), pp. 227–40.

40. Mexico, Secretaría de Fomento, Colonización e Industria, Dirección General de Estadística, *Boletín*, no. 1 (1912), p. 75.

41. Any theory of the Mexican Revolution that depended on such a mechanical linkage between grievances and revolt would have very little utility. This work is not a study of the Mexican Revolution, so elaborating a theory of revolution would be outside its intended scope. Some of the complexities of the matter are considered by Charles Tilly, *From Mobilization to Revolution* (Reading,

Mass.: Addison-Wesley, 1978). Also see John Tutino, *From Insurrection to Revolution in Mexico: Social Bases of Agrarian Violence, 1750–1940* (Princeton: Princeton University Press, 1986).

42. In other words, modernized property relations reduced transaction costs. See Douglass C. North, *Structure and Change in Economic History* (New York: W. W. Norton & Co., 1981), pp. 7, 9, 17–19, 37; Douglass C. North and Robert Paul Thomas, *The Rise of the West: A New Economic History* (Cambridge: Cambridge University Press, 1973), p. 8.

43. A survey increased the value of land by as much as 50 percent, as shown in chapter 5.

CHAPTER 5: THE IMPACT OF THE SURVEYS ON LAND CONCENTRATION AND VALUES

1. See the references cited at n. 46 in chapter 1 as well as the analysis in chapter 6.

2. Unfortunately, this method is biased toward the conclusion that survey company land was subdivided, because positive documentation that a survey company kept the land—i.e., that nothing happened to it—was less likely to show up in the deslindes files.

3. ATN. CHIH 1.322/12, pp. 100–101, 103–5, 76–81; SRE to Fomento, Aug. 9, 1927; contract, Pine King Land & Lumber Company with Sec. de Ag. y Fom., Oct. 30, 1930; pp. 33, 124–6, 93–94; memo, May 8, 1928.

4. ATN. CHIH 54653, cuaderno 1, Arellano to Fomento, Feb. 8, 1886; cuaderno 2, pp. 19–20.

5. NOT. Vicente de P. Velasco, Feb. 25, 1888, pp. 202–7.

6. ATN. CHIH 84703, notarial documents of Sept. 21, 1891, and Feb. 3, 1894.

7. ATN. CHIH 75669, notarial document, Sept. 28, 1909; memo, Jan. 30, 1922.

8. ATN. CHIH 3211, Humphrey to Fomento, Jan. 18, 1905.

9. NOT. Rafael Pérez Gallardo (no. 1), Oct. 16, 1908, vol. 54, pp. 200–219.

10. NOT. José C. Vargas (no. 731), Aug. 1, 1884, pp. 13–14.

11. NOT. José C. Vargas, Feb. 20, 1890, pp. 20–23.

12. Ibid.

13. NOT. Jesús Rosete López, Apr. 21, 1910, vol. 17, pp. 150–60.

14. ATN. DUR 76140, titles, Aug. 23, 1890; DUR 76142, Calderón to Fomento, July 29, 1890. One of the titles was issued to José María Calderón, another Durango contractor, as part of the settlement of a conflict between the two over land they were surveying. The recipients of the others may have been landholders with whom the company arranged composiciones. They were Juan Hernández y Marín, Ignacio Ortega, María de Jesús Alvarez, Antonio Hernández y Prado, and Adolfo Díaz.

15. ATN. SON 78712, Macmanus to Fomento, June 20, 1887; toma de razón, with memo of May 25, 1910.

16. ATN. SON 1.71/36, pp. 31–32, 34, 37, 44–47.

17. ATN. SON 1.71/67, pp. 1, 8, 17, 22.

18. ATN. SON 78729, pp. 209–13.

19. ATN. CHIAP 1.71/24, pp. 199–206.

20. NOT. Gil Mariano León, May 7, 1904, vol. 4, pp. 9–35.

21. ATN. CHIAP 1.71/24, pp. 199–206.

22. ATN. CHIAP 1.322/8, Juan A. Navarro to Fomento, Oct. 23, 1891.

23. ATN. CHIAP 1.71/24, p. 60.

24. NOT. José de Jesús Arce (no. 30), Apr. 16, 1906, vol. 10, pp. 84–125.

25. ATN. CHIAP 1179, pp. 53–69.

26. ATN. CHIAP 1179, pp. 116–19.

27. NOT. Rafael Carpio, Mar. 2, 1906, vol. 20, pp. 192–99; Mar. 24, 1906, vol. 19, pp. 205–10; and Jan. 20, 1906, vol. 16, pp. 176–190; Francisco Aguirre de Pino, Dec. 15, 1906, vol. 22, pp. 107–14. Bernardo Romero, Nov. 25, 1909, vol. 25, pp. 134–41. See ATN. CHIAP 78654, memo, Feb. 22, 1960, for a summary of these transactions.

28. ATN. SIN 1.71/72, Francisco de P. Alvarez to Alson J. Streeter, Mar. 25, 1901.

29. ATN. SIN 1.71/40, pp. 55, 67.

30. These prices were set every two years for each state. They are given in De la Maza, *Código,* passim, and Villamar, *Las leyes federales,* pp. 227–40.

31. Baldío was usually sold outright by Fomento, but it could also be acquired through the composición procedure, by prescription, or by *"labradores pobres"* (poor farmers) under a program designed to assist poor peasants. In none of these instances was a free market operating.

32. ATN. DUR 20, pp. 43–47.

33. ATN. SIN 1.71/20, memos, Jan. 10, 1894, and Dec. 28, 1893.

34. ATN. SON 1.71/64, p. 86; SIN 1.71/40, pp. 29–48; TAB 1.71/37, memo, June 13, 1917.

35. ATN. CHIAP 1.71/24, pp. 244–48.

36. ATN. SIN 113022, p. 30; 1.71/154, p. 37; 1.71/163, p. 28; 1.71/169, p. 36; 1.71/170, p. 34; 1.71/173, p. 46; 1.71/99, p. 48.

37. ATN. SIN 1.71/124, p. 43; SIN 1.71/125, p. 52; SIN 113034, Fomento to Zevada, June 11, 1907.

38. In the case of General Olivares's concession, one of two partners obtained title to all the land surveyed, then transferred all of it to the other partner. It seems likely that the well-placed general was using his influence to benefit a foreigner whose access to the government was more restricted. The land involved bordered the United States, and because Olivares's partner, John Beatty, was probably a U.S. citizen, a title issued directly in Beatty's name would have appeared imprudent at best.

39. The secondary and primary sources (largely state government reports) regarding the six states during the late nineteenth century and early twentieth century were researched in the Biblioteca Nacional in Mexico City. The literature was searched for the names of survey contractors active in those states, but only in one case was a specific reference found to large landholdings by a surveyor.

Policarpo Valenzuela of Tabasco was said to possess fincas totaling more than 200,000 hectares. See Salvador Teuffer, *El Departamento Agrario en la integración de Tabasco a la revolución mexicana* (Mexico: n.p., 1935), p. 26.

CHAPTER 6: THE SURVEY COMPANIES AND THE REVOLUTION OF 1910

1. Mexico, Sec. de Fomento, *Memoria*, 1901–4, p. xii; a copy of the decree can be found at pp. 5–7 among the anexos.

2. Covarrubias, *Varios informes*, pp. 10–15, 6.

3. For the gradual shift in public land policy that started in the mid-1890s toward protecting the occupants of public land, see Donald Fithian Stevens, "Agrarian Policy and Instability in Porfirian Mexico," *Americas* 34 (October 1982): 162–63; Phipps, *Aspects*, pp. 128–29; Cosío Villegas, *La vida social*, p. 193.

4. Thomas Benjamin and Marcial Ocasio-Meléndez, "Organizing the Memory of Modern Mexico: Porfirian Historiography in Perspective, 1880s–1980s," *Hispanic American Historical Review* 64 (May 1984): 337.

5. Orozco, *Legislación y jurisprudencia*, pp. 929–30; 825; 827–28.

6. Ibid., p. 800.

7. Ibid., p. 914.

8. Ibid., pp. 917–19.

9. Ibid., p. 926.

10. Ibid., p. 967.

11. See SRE. 14–4–10, p. 16. Also, CPD. 41/7.15/94–96; 14/3/1343–44; 14/8/3865–66, 3875–86; 41/7.15/216–18.

12. Gavin Kitching, *Development and Underdevelopment in Historical Perspective: Populism, Nationalism and Industrialization* (London: Routledge, 1989), ch. 2. Of course, Orozco's critique of large estates and his arguments in favor of encouraging a class of smallholders were not novel in Mexico; see the fifth chapter of Enrique Florescano's *Origen y desarrollo de los problemas agrarios de México, 1500–1821* (Mexico: Ediciones Era, 1976), which notes that the main opposition to the large estates originated from within the colonial oligarchy itself in the late eighteenth century.

13. D. A. Brading, "Introduction: National Politics and the Populist Tradition," in *Caudillo and Peasant in the Mexican Revolution*, ed. D. A. Brading, (Cambridge: Cambridge University Press, 1980), p. 14.

14. Ibid.

15. In his chapter on landed property, Moliná Enríquez does not even mention the survey companies—a curious omission.

16. Frank Tannenbaum, *Peace by Revolution: An Interpretation of Mexico* (New York: Columbia University Press, 1933; *Peace by Revolution: Mexico after 1910*, New York: Columbia Paperback Edition, 1966), p. 142.

17. Phipps, *Aspects*, p. 115; McBride, *Land Systems*, pp. 73–74.

18. Orozco, *Legislación y jurisprudencia*, p. 895.

19. Enrique Florescano, "La influencia del estado en la historiografía mex-

icana," *Siempre!,* Aug. 18, 1976, pp. x–xi.

20. In addition to the works by Friedrich, Falcón, Hernández Chávez, and Wasserman cited in nn. 21–24, the other studies examined are Frans J. Schryer, *The Rancheros of Pisaflores: The History of a Peasant Bourgeoisie in Twentieth-Century Mexico* (Toronto: University of Toronto Press, 1980); Ian Jacobs, *Ranchero Revolt: The Mexican Revolution in Guerrero* (Austin: University of Texas Press, 1982); Margarita Menegus Bornemann, "Ocoyoacac—Una comunidad agraria en el siglo XIX," *Historia mexicana* 30 (July–September 1980): 33–78; Charles R. Berry, *The Reform in Oaxaca, 1856–76: A Microhistory of the Liberal Revolution* (Lincoln: University of Nebraska Press, 1981); Raymond Th. J. Buve, "Peasant Movements, Caudillos and Landreform [*sic*] during the Revolution (1910–1917) in Tlaxcala, Mexico," *Boletín de estudios latinoamericanos y del Caribe* 18 (June 1975): 112–52; John Womack, Jr., *Zapata and the Mexican Revolution* (New York: Vintage Books, 1970); Allen Wells, "Family Elites in a Boom-and-Bust Economy: The Molinas and Peóns of Porfirian Yucatán," *Hispanic American Historical Review* 62 (May 1982): 224–53; G. M. Joseph, *Revolution from Without: Yucatán, Mexico and the United States, 1880–1924* (Cambridge: Cambridge University Press, 1982); Thomas Benjamin, "Revolución interrumpida—Chiapas y el interinato presidencial—1911," *Historia mexicana* 30 (July–September 1980): 79–98; Evelyn Hu-Dehart, "Pacification of the Yaquis in the Late Porfiriato: Development and Implications," *Hispanic American Historical Review* 54 (February 1974): 337–51; Hector Aguilar Camín, *La frontera nómada: Sonora y la revolución mexicana* (Mexico: Siglo Veintiuno Editores, 1977).

21. Paul Friedrich, *Agrarian Revolt in a Mexican Village* (Chicago: University of Chicago Press, 1977), pp. 44, 46.

22. Romana Falcón, "¿Los orígenes populares de la revolución de 1910? El caso de San Luis Potosí," *Historia mexicana* 29 (October–December 1979): 206–8.

23. Alicia Hernández Chávez, "La defensa de los finqueros en Chiapas—1914–1920," *Historia mexicana* 28 (January–March 1979): 346–47, 351.

24. Wasserman, *Capitalists, Caciques, and Revolution,* pp. 107–8. Wasserman's claim that the Tomochí revolt was partly the result of survey company activity is not supported by Francisco Almada, whose *La rebelión de Tomochí* (Chihuahua, Sociedad Chihuahuense de Estudios Históricos, 1938) Wasserman cites. Almada's later *La revolución en el estado de Chihuahua,* 2 vols. (Chihuahua: Biblioteca del Inst. Nacional de Estudios Históricos de la Revolución Mexicana, 1964), 1:99–103, does not mention the survey companies and attributes the revolt to interference by outside authorities. (The latter work's eighth chapter identifies thirteen separate outbreaks of "inconformidad o rebeldía" against the regime before 1910, but in none does the author mention survey company activities.) Alan Knight, "Peasant and Caudillo in Revolutionary Mexico, 1910–1917," in *Caudillo and Peasant,* ed. D. A. Brading, p. 28, also interprets the Tomochí revolt this way. Wasserman himself elsewhere (p. 40) called Tomochí a "religious uprising."

25. Friedrich Katz, "Mexico: Restored Republic and Porfiriato, 1867–1910," in *The Cambridge Modern History of Latin America,* ed. Leslie Bethell (Cam-

bridge: Cambridge University Press, 1984–), 5:70. Further evidence of a post–1900 "científico offensive" was offered by Wasserman, *Capitalists, Caciques, and Revolution,* p. 109, who states that the revolution in Chihuahua in 1910 was touched off by the implementation of the state's Municipal Land Law of 1905, which empowered "latifundists and speculators . . . to appropriate the lands of small holders, villages, and municipalities."

26. In 1963 a forest-products company asked the government for the maps of Cantón Degollado made by Jesús Valenzuela's company in 1884, to interpret the boundary agreements between the states of Chihuahua and Sonora. The next year a farmers' organization asked for the same maps to identify land requested by *ejidatarios* (holders of ejidos). ATN. CHIH 54653, cuaderno 2, Ing. Emilio Flores Calderón to Dirección de Terrenos Nacionales, Jan. 2, 1963; Liga de Comunidades Agrarias y Sindicatos del Estado de Sonora to Sec. de Ag. y Ganadería, July 19, 1962.

27. ATN. DUR 1.71/24, pp. 85–87.

28. Covarrubias, *Varios informes,* pp. 27–28.

29. Ibid., pp. 29–32.

30. Villamar, *Las leyes federales,* appendix, pp. 3–23; Mexico, Sec. de Fomento, *Memoria,* 1909–10, p. v; Mexico, Sec. de Fomento, *Memoria,* 1910–11, pp. lxi–lxii.

31. Francisco I. Madero, "Plan de San Luis Potosí," in *México en el siglo XX, 1900–1913: Textos y documentos,* eds. Mario Contreras and Jesús Tamayo (Mexico: Universidad Nacional Autónoma de Mexico, 1975), pp. 327–28. The outlawed Partido Liberal Mexicano had called four years earlier for united opposition to the regime under its program, which did not specifically criticize Porfirian land policy. Less radical than Madero's plan, it merely promised to force landowners "hacer productivos todos sus terrenos, so pena de perderlos." Rather vaguely and incoherently, it then added that the restitution of ejidos to pueblos that had lost them "es de clara justicia." See "Programa del Partido Liberal Mexicano y Manifiesto a la Nación" in the aforementioned collection, pp. 241–42.

32. "Ley de 6 de enero de 1915," in *México en el siglo XX, 1913–1920: Textos y documentos,* eds. Mario Contreras and Jesús Tamayo (Mexico: Universidad Nacional Autónoma de México, 1976), pp. 180–83.

33. "Artículo 27 de la Constitución de 1917," in Contreres and Tamayo, eds., *Mexico, 1913–20,* pp. 262–67.

34. The nullification of titles was limited to the one-third granted in compensation; the two-thirds, if alienated, had been sold directly by the government and were not affected by this provision of Article 27.

35. ATN. CHIH 1.322/12, pp. 42–62.

36. *Diario Oficial,* Jan. 14, 1925, pp. 254–56.

37. Luis Miguel Díaz, ed., *México y las comisiones internacionales de reclamaciones,* 2 vols. (Mexico: Universidad Nacional Autónoma de Mexico, 1983), 1:187.

38. ATN. CHIH 1.322/12, contract, Oct. 30, 1930.

39. ATN. CHIH 1.322/12, memo, May 8, 1928; presidential decrees, Sept. 13, 1928, and Aug. 21, 1930.

40. ATN. CHIH 1.322/12, Ing. Martín Rosales M. to Sec. Gral. of CRC,

Jan. 27, 1958; CRC to Agencia General de Agricultura y Ganadería, Feb. 1, 1958; memo, Sept. 14, 1972.

41. See, for example, ATN. CHIH 8245, Gerardo Calderón et al. to Sec. de Agric. y Fomento, May 20, 1924.

42. ATN. TAB. 1.71/4, "Informe," Jan. 31, 1916; 1.71/3, pp. 54–55; *Diario Oficial,* Sept. 13, 1917.

43. ATN. TAB 1.71/32, memo, Feb. 27, 1930; Teuffer, *Departamento Agrario,* p. 26.

44. ATN. CHIH 8245, "Acuerdo presidencial," June 26, 1924, and Oct. 4, 1928.

45. *Diario Oficial,* Feb. 7, 1939, pp. 3–5.

46. ATN. SIN 84792, "Extracto," Dec. 4, 1923.

47. Positive evidence that these lands were untouched can be found in correspondence between the governor of Sinaloa and the Sec. de Ag. y Ganadería in 1952. See ATN. SIN 84793, pp. 130–31, and ATN. CHIAP 78654, memo, Nov. 28, 1953. Some of the land was auctioned by the state for back taxes. See SIN 84793, pp. 130–31, and SIN 84868, pp. 210–18.

48. ATN. DUR [*sic*] 9053, cuaderno 3, "Relación de los Terrenos Nacionales situados en el Estado de Sinaloa, que le fueron titulados de Luis Martínez de Castro," Dec. 13, 1954.

49. ATN. SIN 1.71/188, "Informe," Aug. 31, 1922.

50. ATN. SIN 1.71/152, Fernando Estrada to Sec. de Ag. y Fomento, Apr. 12, 1926; reply, May 20, 1926.

51. ATN. CHIAP 78654, memo, Sept. 25, 1914; Martínez de Castro to Sec. de Ag., Oct. 16, 1920; Sec. de Ag. to suc. Martínez de Castro, Oct. 29, 1921; contract, Feb. 28, 1922; Antonio Martínez de Castro to Sec. de Ag., Mar. 16, 1927.

52. This company contracted with the government to exploit timber and buy public lands in Chiapas, but it never held a survey contract. Its authority to survey this parcel was issued rather informally (and perhaps illegally). No other surveys appear to have been conducted on such grounds. Because of its unique status, the company was not counted as a survey contractor in the data summaries.

53. ATN. CHIAP 91000, cuaderno 1, pp. 440–41, 460–67; cuaderno 2, pp. 73–77; cuaderno 3, pp. 8–11.

54. *Diario Oficial,* Feb. 19, 1923; ATN. CHIAP 84797, memo, Sept. 29, 1923. In March 1917, President Carranza, reaching back to the 1883 colonization law for legal authority, nullified some of the titles because of the company's failure to fulfill the obligations of its colonization contracts. (See *Diario Oficial,* Apr. 3, 1917, pp. 373–74.) Negotiations on an indemnity began as early as 1919, however. (See ATN. CHIAP 1179, pp. 212–13.) The Obregón administration decree providing for the indemnity also noted that it had asked the Cámara de Diputados to set aside $50 million to pay the British companies "y las otras Compañías" whose concessions were revoked in 1917. The British companies' lawyer delivered inventories and other documents relating to the land in question to the government in 1927 (ATN. CHIAP 1179, p. 273).

55. *Diario Oficial,* Oct. 27, 1930, sect. 2A, pp. 1–2.

56. ATN. CHIAP 1179, Sec. de Ag. to José Quiyono Bolaños, Sept. 29, 1939.

57. ATN. CHIAP [*sic*] 1179, "Relación de expedientes de reconocimientos de propiedad derivados de la concesión Ignacio Gómez del Campo y R. Guerrero." Guerrero."

58. ATN. CHIAP [*sic*] 1179, "Relación de expedientes de reconocimientos de propiedad derivados de la concesión de Jesús E. Valenzuela."

59. ATN. CHIAP [*sic*] 1179, "Relación de expedientes de reconocimientos de propiedad derivados de la concesión de Rafael García Martínez."

60. Leopoldo Solís, *La realidad económica mexicana: Retrovisión y perspectivas*, rev. ed. (Mexico: Siglo Veintiuno Editores, 1981), pp. 79–84.

61. Córdova, *Formación del poder político*, pp. 31–33.

62. Katz, *Secret War*, p. 321.

63. Linda B. Hall, *Alvaro Obregón: Power and Revolution in Mexico, 1911–1920* (College Station: Texas A&M University Press, 1981), pp. 251–52.

64. Contreras and Tamayo, *Mexico en el siglo XX: 1913–1920*, pp. 372–78.

65. Hall, *Alvaro Obregón*, pp. 257–58.

66. Simpson, *Ejido*, p. 87.

67. Rafael Loyola Díaz, *La crisis Obregón-Calles y el estado mexicano* (Mexico: Siglo Veintiuno Editores, 1980), p. 14. Hall, *Alvaro Obregón*, p. 255, credits the Sonoran with contributing significantly to overcoming this fragmentation and centering power in the presidency, a task that Calles would carry even further.

68. Robert Freeman Smith, *The United States and Revolutionary Nationalism in Mexico, 1916–1932* (Chicago: University of Chicago Press, 1972), pp. x–xi, 162, 175, 178, 190, 194, 195, 200.

69. United States–Mexican Commission, *Proceedings of the United States–Mexican Commission Convened in Mexico City, May 14, 1923* (Washington: U.S. Government Printing Office, 1925), p. 26.

70. Smith, *United States*, pp. 221–23.

71. Titles issued to Luis Huller and the International Company are counted as having been revoked because they were later acquired by the Mexican Land Colonization Company, whose titles were subsequently nullified.

72. "Apparently" is a necessary modifier in this case because it was only the absence of evidence of revocation that suggests the conclusion. This was true for all those contractors not listed in Table 16 except for Luis Martínez de Castro and the Sinaloa Land Company, as noted earlier. It is possible, therefore, that decrees affecting those companies not listed in Table 16 (except for the two Sinaloa contractors) were also issued.

73. Lorenzo Meyer, "Historical Roots of the Authoritarian State in Mexico," in *Authoritarianism in Mexico,* ed. José Luis Reyna and Richard S. Weinert (Philadelphia: Institute for the Study of Human Issues, 1977), p. 4.

CHAPTER 7: SUMMARY AND CONCLUSIONS

1. "Surveying companies were notorious for usurping village common lands," reports Wasserman, *Capitalists, Caciques, and Revolution*, p. 50. He mentions

two cases. The first (p. 50), citing a newspaper report, states that a survey company "caused widespread agitation among the region's Tarahumara Indians." The second, which claims that survey company activities were partly to blame for the revolt of the village of Tomochí in 1892 (p. 108), is not even backed up by the source that the author cites, nor by the author himself elsewhere in his book (p. 40). In *El proceso de modernización capitalista en el noroeste de Chihuahua (1880–1910)* (Mexico: Universidad Iberoamericana, Departamento de Historia, 1987), Jane Dale-Lloyd writes that surveys by three companies in Chihuahua "fueron muy irregular; en un sinnúmero de casos afectaron las tierras de los antiguos ejidos de los pueblos del área, y en no pocas ocasiones a pequeños propietarios" ("were very irregular; in a huge number of cases they affected the land of the former ejidos of the pueblos of the area, and in not a few occasions, small holders") (p. 72). While no cases of expropriations are actually cited by the author, she quotes a newspaper report that three villages that together claimed 112,359 hectares of ejido on the basis of an eighteenth-century grant were left with only 28,080 hectares following the surveys in 1885 (p. 74). Apart from the limitations of the source and the failure to develop the nature of the "irregularity" of the survey, the author fails to consider the vagueness and unreliability of colonial-era titles.

2. See the discussion in chapter 6. Critiques of the agrarian interpretation can be found in D. A. Brading's introduction to *Caudillo and Peasant*, and in Jean-A. Meyer, "Le Mexique a la veille de la révolution de 1910: Crédibilité des statistiques agraires," *Revue historique* 562 (April–June 1987). Jan Bazant points out cases of hacienda disintegration in the nineteenth century that benefited smallholders, in "La división de las grandes propiedades rurales mexicanas en el siglo XIX," in *Despúes de los latifundios (la desintegración de la gran propiedad agraria en México)*, coord. Heriberto Moreno García (N.p.: Colegio de Michoacán, 1982), pp. 33–48. A similar trend well into the Porfiriato was documented by Héctor Díaz-Polanco and Laurent Guye Montandon, *Agricultura y sociedad en el Bajío (S. XIX)* (Mexico: Centro de Investigación para la Integración Social, 1984), pp. 91–95. Tutino, *Insurrection to Revolution,* partially attributes revolutionary grievances to land usurpation in the northern frontier zone by entrepreneurs and by the state government (pp. 299–300), and in the central highlands among villagers whose community land was privatized by the Díaz state (p. 318). Tutino notes that demographic change, economic downturns, and changes in the structure of agricultural production and in social relations of production also contributed to revolutionary action.

3. Roger D. Hansen, *The Politics of Mexican Development* (Baltimore: Johns Hopkins University Press, 1971), p. 15.

4. William B. Taylor, *Drinking, Homicide and Rebellion in Colonial Mexican Villages* (Stanford: Stanford University Press, 1979), p. 166.

5. Charles Gibson, *The Aztecs under Spanish Rule: A History of the Indians of the Valley of Mexico, 1519–1810* (Stanford: Stanford University Press, 1964), p. 271.

6. Ibid., pp. 287–88; Taylor, *Drinking, Homicide and Rebellion,* p. 170.

7. Frans J. Schryer, "Peasants and the Law: A History of Land Tenure and Conflict in the Huasteca," *Journal of Latin American Studies* 18 (November 1986) 2:308–11.

8. E. P. Thompson, *Whigs and Hunters: The Origin of the Black Act* (New York: Pantheon Books, 1975), pp. 258–63.

9. See chapter 1.

10. Adolfo Gilly, "La guerra de clases en la revolución mexicana (Revolución permanente y auto-organización de las masas)," in *Interpretaciones de la revolución mexicana* (Mexico: Editorial Nueva Imagen, 1980), p. 48.

11. Tutino, *Insurrection to Revolution*, p. 282.

12. Katz, "Mexico," pp. 68–70; and Tutino, *Insurrection to Revolution*, p. 332, have noted that by 1900 the state was forced to yield to certain factions of the ruling class some of the autonomy it had earned earlier. For more evidence of the regime's flexibility in dealing with labor and peasant grievances, see Stevens, "Agrarian Policy and Instability," pp. 153–66; and David Walker, "Porfirian Labor Politics: Working Class Organizations in Mexico City and Porfirio Díaz, 1876–1902," *Americas* 37 (January 1981): 257–90.

APPENDIX A: TWENTY SURVEY ENTERPRISES

1. ATN. CHIH 97449, pp. 113–15; DUR 113062, pp. 161–63, 197, 212.

2. Wasserman, *Capitalists, Caciques, and Revolution*, p. 39; Francisco Almada, *Diccionario de historia, geografía y biografía Chihuahuenses* (Chihuahua, Mexico: Talleres Gráficos del Gobierno del Estado, 1927), pp. 82–83.

3. ATN. SIN 84792, pp. 72–74; SIN 1.71/29, Ignacio de la Cueva to President Díaz, Aug. 6, 1889.

4. ATN. SIN 84792, pp. 38–42.

5. NOT. Vicente de P. Velasco (no. 730), Nov. 15, 1889, pp. 489–92; Manuel Alvarez de la Cadena (no. 4 A.H.), Feb. 1, 1890, pp. 74–75.

6. Pablo was a well-known lawyer and financier, especially active in the railroad industry until his death in 1907. His brother Manuel acquired rights to survey in Sonora and Sinaloa. For the transactions with Fomento, see ATN. SIN 84792, pp. 59, 68, 65; SIN 1.71/38, Manuel Martínez del Río to Fomento, Oct. 6, 1891, and Fomento to Martínez del Río, Oct. 16, 1891. Also see NOT. Manuel Alvarez de la Cadena (no. 4 A.H.), Feb. 1, 1890, pp. 76–77.

7. ATN. SIN 1.71/89, MacWood to Fomento, Oct. 24, 1894; SIN 1.71/65, MacWood to Fomento, May 28, 1894; SIN 1.71/68, Martínez del Río to Fomento, Apr. 15, 1899.

8. ATN. SIN 1.71/48, Eduardo Andrade to Agente de Tierras, May 25, 1896.

9. Cosío Villegas, *La vida política interior*, pt. 2, p. 858; Graciela Fix Zamudio, "Joaquín D. Casasús: Humanista mexicano del siglo XIX" (Licenciado thesis, Universidad Nacional Autónoma de México, 1963), p. 14; José C. Valadés, *El Porfirismo: Historia de un régimen*, 3 vols. (Mexico: José Porrúa e Hijos, 1941–48; reprint ed. of vols. 2 and 3, Mexico: Universidad Nacional Autónoma de México, 1977), *El crecimiento: I*, 2:221.

10. ATN. DUR 76140, title, Aug. 23, 1890. One of the titles was issued to José María Calderón, another Durango contractor, as part of the settlement of a conflict between the two over land they were surveying. The recipients of the

others may have been landholders with whom the company arranged composiciones. They were Juan Hernández y Marín, Ignacio Ortega, María de Jesús Alvarez, Antonio Hernández y Prado, and Adolfo Díaz (DUR 76142, Calderón to Fomento, July 29, 1890).

11. ATN. CHIH 3216, p. 58.

12. ATN. DUR 1.71/6, pp. 1–3. Rafael García Martínez had also begun surveying in Durango in 1884, but his company was not mentioned in any correspondence or other documents connected with the merger.

13. CPD. 7/2/476–78. Díaz may have shed the stock before he was reinaugurated as president in December 1884. Offered shares in a Guerrero silver mine by Gen. Canuto A. Neri two weeks after his reinauguration, Díaz declined, because "from the moment of my inauguration as President, I cannot nor do I want to appear as a partner in any private business, because besides the fact that it would not be proper, there is the circumstance that if for whatever reason and in the course of time should any difficulty be stirred up among the partners or the businesses, I would not be free and independent to proceed with all the rectitude with which I am accustomed to do" (CPD 9/3/1021).

14. Cosío Villegas, *La vida política interior,* pt. 1, p. 751.

15. CPD. 7/2/476–478. Pacheco, Ceballos, and Romero Rubio were also partners in a gold mine claim in Baja California in 1883, along with Francisco Cañedo (the governor of Sinaloa and a senator from Sonora in the 1880s), Gen. José Guillermo Carbó, and others. See CPD 8/1/82.

16. ATN. DUR 1.322/17, Meza to Fomento, Apr. 22, 1883. General Meza himself was not on the board, though with eight shares he was the biggest investor. See CPD 7/2/476–78.

17. ATN. DUR 1.322/17, Romero Rubio to Fomento, June 11, 1884.

18. NOT. José María Velásquez (no. 732) July 25, 1884, pp. 16–18.

19. NOT. José María Velásquez (no. 732), July 25, 1884, pp. 18–24; José C. Vargas (no. 731), July 29, 1884, pp. 12–13; José María Velásquez, Aug. 12, 1885, pp. 166–69.

20. ATN. DUR 113062, pp. 197, 212. NOT. Agustín Pérez de Lara (no. 62), Oct. 2, 1906, vol. 33, pp. 183–91. This instrument indicates that when Pacheco's widow, Josefa Calderón, needed to raise some cash in 1906, she sold the general's three shares to the Banco de Londres y México and Asúnsolo's estate for $3,440. This price was calculated by figuring that the number of hectares corresponding to three shares was about 2,295; each hectare was then valued at $1.50. For three shares to represent 2,295 hectares, there must have been 455 shareholders by then.

21. Originally the twenty had been distributed thus: Quaglia (two), Lancaster Jones (five), Poucel (four), Rocha (four), and Escoto (five). NOT. Agustín Pérez de Lara (no. 62), Mar. 7, 1907, vol. 41, pp. 163–77; ATN. DUR 113062, pp. 197, 212.

22. In an allusion to Hammeken's cover-up for Díaz and Pacheco, the notarial document states that even though the company charter gave him nine shares, his will said he only owned three. NOT. Agustín Pérez de Lara (no. 62), July 29, 1907, vol. 45, pp. 146–51.

23. Cosío Villegas, *La vida política interior,* pt. 1, pp. 428–29, 578; pt. 2, pp. 25, 279.

24. ATN. SIN 84794, García et al. to Fomento, May 21, 1883.

25. ATN. CHIH 97449, pp. 60, 113–15; CHIH 54653, cuaderno 1, court file, survey of Cantón Meoqui; CHIH 77415, pp. 59-61; SON 78729, pp. 209–13.

26. ATN. CHIH 54653, cuaderno 2, p. 30.

27. ATN. CHIH 97322.

28. Cosío Villegas, *La vida política interior,* pt. 2, p. 680; CPD 9/1/178, 196.

29. Ireneo Paz, *México actual: Galería de contemporaneos* (Mexico: Ofc. Tip. de la Patria, 1898), pp. 57–58.

30. NOT. José C. Vargas, Feb. 20, 1890, pp. 20–23.

31. Ibid.

32. NOT. José C. Vargas, Mar. 10, 1890, pp. 30–42.

33. Ibid.

34. ATN. DUR 111125, pp. 77–81.

35. Ibid.

36. ATN. CHIH 84771, Fomento to Patricio Gómez del Campo, Dec. 11, 1886.

37. ATN. CHIH 97449, pp. 64–66; CHIH 119, Macmanus et al. to Juez del Distrito, Jan. 2, 1883; CHIH 84771, Patricio Gómez del Campo to Fomento, Mar. 17, 1883. Juan Terrazas, one of the sons of Chihuahua caudillo Luis Terrazas, apparently acquired a stake in the company later on. See CHIH 1.71/120, Gómez del Campo to Fomento, Apr. 4, 1898.

38. CPD. 41/5.12/143–46, 176–78, 246–50, 330, 443–45; 41/6.13/108–12; 11/14/6548–49.

39. Mexico, Secretaría de Fomento, *Boletín semestral de la estadística de la república mexicana,* no. 3 (1889):209–11.

40. The source for this paragraph sometimes specified "pesos" and other times only gave the $ sign. In this case, because of the ambiguity of the source, the sign ?$ is used.

41. SRE. 15–28–2, Romero to SRE, Dec. 24, 1888; 15–28–2, Romero to SRE, Mar. 8, 1889, and Fomento to SRE, May 17, 1889. A copy of the prospectus can be found in SRE 3743–15, pp. 83–87.

42. CPD. 14/10/4775–78; SRE 15–28–2, Romero to SRE, May 31, 1889; Romero to Wells, June 5, 1889; Wells to Romero, June 6, 1889. See also an article from the *New York Herald,* May 31, 1889, in SRE 15–28–2.

43. A copy of the charter is in ATN. CHIAP 1.71/22, pp. 32–93.

44. ATN. CHIAP 1.71/22, pp. 76–86.

45. AGN. Fondo Vera Estañol, 15/12/1. A copy of Huller's will is in this collection. ATN. CHIAP 84797, Huller to Fomento, Sept. 11, 1928; Fomento to Huller, Sept. 22, 1928.

46. NOT. José de Jesús Arce (no. 30), Apr. 16, 1906, vol. 10, pp. 84–125; ATN. CHIAP 1.71/24, pp. 363–66.

47. ATN. SIN 1.71/29, Ignacio de la Cueva to President Díaz, Aug. 6, 1889; SIN 1.71/206, which includes notarial document dated Feb. 6, 1894,

executed before Notary Jesús Raz Guzmán.

48. ATN. SIN 84793, pp. 107, 111; CHIAP 78654, memo, Sept. 25, 1914.
49. NOT. Heriberto Molina, Jan. 27, 1905, vol. 7, pp. 26–31.
50. See the "contrato de sociedad" in ATN. SIN 84868, pp. 50–54, and amendments at pp. 57–79.
51. AGN. Fondo Vera Estañol, 24/4/1, "Testamento público y abierto de la Señora Mercedes Esqueda de Martínez de Castro."
52. ATN. SIN 84868, p. 49.
53. CPD. 10/15/7275–76; 10/23/11100; 10/24/11641–42. Valadés, *El crecimiento: I,* 2:241; ATN. SON 1.71/36, pp. 31–32, 34, 37, 44–47.
54. ATN. DUR 1.71/32, Juez de Distrito to Fomento, Oct. 23, 1888; SON 78712, court file.
55. NOT. Rafael F. Morales (no. 444), June 1, 1885, pp. 247–57.
56. Ireneo Paz, *Los hombres prominentes de México* (Mexico: La Patria, 1888), pp. 309–10. NOT. José María Velásquez (no. 732), July 31, 1885, pp. 155–57; ATN. SON 97554, pp. 179–82; SON 1.322/12, memo, Feb. 16, 1893. The contract was published in De la Maza, *Código,* pp. 999–1011.
57. ATN. SON 1.322/12, passim; CHIAP 1.71/24, pp. 199–206, 327–35.
58. ATN. SIN 1.71/88, Ignacio F. de Alfaro to Fomento, Dec. 23, 1893; SIN 1.71/206, notarial document, Feb. 6, 1894, before Jesús Raz Guzmán of Mexico City.
59. ATN. CHIH 1.322/12, pp. 100–101.
60. ATN. CHIH 54653, cuaderno 2, p. 30.
61. Joaquín Márquez Montiel, *Hombres célebres de Chihuahua* (Mexico: Editorial Jus, 1953), pp. 90–91; CPD. 10/16/7924–25; 10/18/8730–31; 11/9/4388–89.
62. ATN. Vicente de P. Velasco (no. 730), Jan. 19, 1883, pp. 51–55; NOT. Vicente de P. Velasco (no. 730), May 29, 1885, pp. 580–82, and July 13, 1885, pp. 23–27.
63. NOT. Vicente de P. Velasco, Feb. 25, 1888, pp. 202–7.
64. ATN. SON 1.71/39, Valenzuela to Fomento, May 11, 1889; SON 1.71/6, García to Fomento, July 25, 1884; SIN 89051, court file (pp. 48–66); CHIH 97449, p. 59; CHIH 1.322/13, pp. 17–29.
65. *Los hombres del centenario* (Mexico: n.p., 1910), 1:112–13; Manuel González Calzada, *Historia de la revolución mexicana en Tabasco* (Mexico: Consejo Editorial del Gobierno del Estado de Tabasco, 1981), pp. 71–72; Manuel Mestre Ghigliazza, *Apuntes para una relación cronológica de los gobernantes de Tabasco* (Mérida: n.p., 1934), p. 155.
66. CPD. 12/6/2568–69, 11/13/6026–27, 41/3.7/377–78.

APPENDIX B: ALLEGATIONS OF LAND USURPATION IN THE SIX STATES

1. ATN. CHIH, 1.71/119, pp. 32–34, 38–40.
2. ATN. CHIH, 1.71/56/134–35.
3. ATN. CHIH, 97374, p. 40.

4. Ibid.

5. ATN. CHIH 97374, pp. 42–43; 1.71/56, pp. 186–88.

6. ATN. CHIH 1.71/111, Pedro Carrillo to Fomento, Apr. 20, 1886; Carlos Pacheco to Cia. Gómez del Campo, Apr. 27, 1886.

7. Almada, *Revolución*, 1:18.

8. ATN. CHIH 1.71/111, Pacheco to Cia. Gómez del Campo, Apr. 27, 1886.

9. ATN. CHIH 1.71/56, pp. 215–21, 228–30, 234.

10. Ibid., pp. 215–21.

11. Ibid.

12. Ibid., p. 226, 230, 233.

13. Ibid., pp. 224–25, 232, 244.

14. ATN. CHIH 1.71/120, memo, Nov. 16, 1896.

15. ATN. CHIH 93237, José Diego Fernández to Fomento, Apr. 8, 1890; Fomento to José Valenzuela, Apr. 11, 1890; Fomento to Hacienda, Apr. 16, 1890; titles, May 15, 1890 and Feb. 2, 1898; CHIH 8245, memo, July 4, 1923.

16. ATN. CHIH 84703, Jefe Político to Fomento, Aug. 22, 1897.

17. The area, formerly in Cantón Matamoros, became part of Distrito Arteaga when the state was redistricted between 1887 and 1893.

18. ATN. CHIH 1.24/15, pp. 1–11.

19. See a reference to this title in ATN. CHIH 84703, Valenzuela to Fomento, Sept. 9, 1894.

20. ATN. CHIH, 78534, pp. 45–47.

21. Ibid., pp. 49–58.

22. Ibid., nine *agricultores* (farmers) to Sec., Nov. 2, 1920.

23. ATN. CHIH 1.29/88, pp. 1–88.

24. ATN. CHIH, 1.322/13, pp. 11–13, 16.

25. ATN. CHIH 1.71/10, court file; memo, Dec. 5, 1885; title, Dec. 13, 1885.

26. ATN. CHIH 54653, cuaderno 2, pp. 27–28, and cuaderno 1, Valenzuela to Fomento, July 31, 1886; CHIH 3216, memos of Sept. 20 and Oct. 5, 1886; Valenzuela to Fomento, July 31, 1886.

27. ATN. CHIH 1.71/20, Falomir to Fomento, May 30, 1885; memo, July 20, 1885; Fomento to Valenzuela, July 23, 1885; Valenzuela to Fomento, July 29.

28. ATN. CHIH 3211, court file.

29. ATN. CHIH 54653, cuaderno 1, memo, Jan. 3, 1885.

30. ATN. CHIH 75669, pp. 1–4; memo, July 8, 1885; Fomento to Juez, Sept. 19, 1885.

31. ATN. CHIH 1.71/21, Hilario Hinojos and Juan Carpintero to Fomento, Mar. 20, 1889; memo, April 23.

32. ATN. CHIH 75669, "Testimonio de deslinde," Cantón Guerrero.

33. Ibid., Escarcegas to Fomento, Apr. 29, 1889; memo, June 18, 1889.

34. ATN. CHIH 75556, p. 32; CHIH 3216, Juez 1a de Letras del Distrito de Iturbide to Fomento, Sept. 11, 1886; Mariano Horcasitas to Fomento, Sept. 27, 1886; memos, Sept. 20 and Oct. 5, 1886; Fomento to Horcasitas, Oct. 12, 1886.

35. ATN. CHIH 75556, pp. 35–36.
36. ATN. CHIH 54653, cuaderno 2, pp. 35–43.
37. ATN. CHIH 77415, court file.
38. ATN. SON 1.71/29, pp. 3–4.
39. ATN. SIN 89051, p. 59.
40. ATN. SIN 1.71/29, De la Cueva to Díaz, Aug. 6, 1889.
41. Ibid.
42. ATN. SIN 1.71/29, Fomento to Gayou and Juez de Distrito, Aug. 29, 1889; 84792, pp. 72–74.
43. ATN. SIN 1.71/58, Islas to Fomento, Nov. 1, 1892.
44. ATN. SIN 1.71/87, Francisco C. Alcalde to Fomento, Feb. 27, 1893, and Mar. 3, 1893.
45. ATN. SIN 1.71/47, Valenzeula de Saracho to Fomento, Nov. 7, 1893.
46. ATN. SIN 1.71/181, court file, pp. 12–55.
47. Ibid.
48. ATN. SIN 1.71/23, telegram, Santiago María Valenzuela et al. to Senador Mariano Martínez de Castro, Jan. 11, 1888.
49. ATN. SIN 1.71/181, court file.
50. ATN. SIN 1.71/30, Juez de Distrito, Nov. 22, 1889.
51. ATN. SIN 1.71/63, pp. 38–39; 1.71/181, pp. 2–6, 116–119.
52. ATN. SIN 1.71/30, memos, Nov. 3, 1897, and Dec. 2, 1897.
53. ATN. CHIH 78534, pp. 40–41, 49. Also see the section on Chihuahua for the company's activities there.
54. Ibid., pp. 16, 23–24.
55. ATN. SIN 113027, Zárate to Fomento, Apr. 26, 1893.
56. Ibid., various memos, 1893–94; title, Oct. 2, 1894.
57. Ibid., Pablo Apolonio y Atilana Pérez to Fomento, Mar. 15, 1894.
58. ATN. CHIH 78534, pp. 68–70.
59. ATN. SIN 1.71/67, memo, Mar. 14, 1899.
60. ATN. SIN 1.71/59, Sotelo to Fomento, Sept. 9, 1899; reply, Nov. 8; Natividad González to Fomento, Sept. 27, 1899; reply, Oct. 30.
61. ATN. SIN 1.71/67, Sandoval to Fomento, Mar. 21, 1902; memo, Apr. 11, 1902; Fomento to Sandoval, Apr. 12, 1902.
62. ATN. CHIH [sic], 1.322/5, decree by Francisco Guemez, May 14, 1883.
63. All these figures were derived by the author from the roughly two hundred separate files on these surveys.
64. Ibid.
65. ATN. SIN 1.71/109, passim.
66. ATN. SIN 1.71/112, passim.
67. ATN. SIN 1.71/113, pp. 50–51.
68. Ibid., passim.
69. ATN. SIN 1.71/112, pp. 35–36, 44.
70. ATN. SIN 1.71/174, pp. 1, 24, 37, 40.
71. ATN. SIN 1.71/140, pp. 32–33; 1.71/139, pp. 32–33.
72. ATN. SIN 1.71/168, pp. 35–36, 41. Titles to public lands issued by

state authorities, not uncommon before the 1850s, were often discarded later; the federal government insisted on its exclusive right to alienate public land. The law of Dec. 3, 1855, declared all titles issued by state authorities without the permission of the central government to be invalid, and the law continued to be applied during the Díaz era, though not without a considerable amount of debate. See the discussion in Orozco, *Legislación y jurisprudencia,* pp. 198–99, 287–92. Yet Fomento was anything but consistent in this regard; when the Sinaloa Land Company rejected a title issued by the governor of Sinaloa, Fomento overruled the company and ordered the federal judge to respect the property (ATN. SIN 84869, memo, Sept. 26, 1908).

73. This phrase was boilerplate on all federal land titles, but in the view of legal scholar Wistano Luis Orozco, writing in 1895, its use was merely traditional, it being "una frase de rutina que no significa nada especial, ni quita al vendedor ninguna de las obligaciones que le impone la ley" (*Legislacion y jurisprudenia,* p. 384).

74. ATN. SIN 1.71/171, passim.

75. ATN. SIN 96719; 1.71/99; 1.71/121; 1.71/197.

76. ATN. SIN 113006, pp. 25, 32–37.

77. ATN. SIN 113006, pp. 48–53.

78. Ibid., pp. 55–72, 97–113.

79. ATN. SIN 113034, passim.

80. The total population of Pueblo Nuevo, located in the partido of Durango, was given as 5,742 in 1893. See Alfonso Luis Velasco, *Geografía y estadística de la república mexicana,* vol. 1, *Durango* (Mexico: Ofc. Tip. de la Sec. de Fomento, 1893), pp. 122, 124.

81. ATN. DUR 9053, cuaderno 1, p. 25.

82. Ibid.

83. Ibid., pp. 25–29.

84. Ibid., p. 30.

85. ATN. DUR 113058, court file; García Martínez to Fomento, Feb. 27 and May 4, 1889; Fomento to García Martínez, May 15, 1889; title, July 24, 1889.

86. ATN. DUR 113070, court file, pp. 4–84, 89–96, 114–115; 113060, Tomás Núñez and José Mata to Fomento, July 30, 1893.

87. ATN. DUR 113060, José Valenzuela and J. Ramón Fernández to Fomento, Nov. 3, 1894; 1.71/20, pp. 18, 21–23; 113070, pp. 101–3, 108–10.

88. ATN. DUR 1.71/20, pp. 32–47.

89. ATN. DUR 1.71/24, pp. 39–40, 46–47, 61.

90. Ibid., pp. 67–72, 75.

91. ATN. DUR 113070, pp. 130, 138–142; 111125, pp. 84–85, 161.

92. ATN. DUR 113062, court file, pp. 5–48.

93. ATN. SON 78712, court file.

94. Ibid., Macmanus to Fomento, Dec. 2, 1886.

95. Ibid., Fomento to Elías, Dec. 10, 1886; correspondence, MacManus and Fomento, Jan.–Feb. 1887; title, Feb. 18, 1887.

96. ATN. SON 1.71/14, Olivares to Fomento, Dec. 26, 1888.

97. ATN. SON 1.71/64, pp. 67–71, 75, 81, 86, 89–90.

98. ATN. CHIAP 1.71/24, pp. 14–16.

99. ATN. CHIAP 1.71/3, pp. 6–10.

100. Ibid., pp. 11–16.

101. ATN. TAB 1.322/2, pp. 130–33, 177.

102. ATN. TAB 1.71/11, court file.

103. ATN. TAB 1.71/24, Moisés Morales and Tomás González to Fomento, Oct. 20, 1913.

104. ATN. TAB 1.71/3, pp. 21, 54–55.

Works Cited

MANUSCRIPT SOURCES

ATN.—Archivo de Terrenos Nacionales. Secretaría de Reforma Agraria, Mexico City.
NOT.—Archivo de Notarías del Distrito Federal, Mexico City.
CPD.—Colección General Porfirio Díaz, Universidad Iberoamericana, Mexico City.
AGN.—Archivo General de la Nación, Mexico City.
SRE.—Archivo Histórico, Secretaría de Relaciones Exteriores, Mexico City.

PUBLISHED SOURCES

Aguilar Camín, Hector. *La frontera nómada: Sonora y la revolución mexicana.* Mexico: Siglo Veintiuno Editores, 1977.
Aguilera Gómez, Manuel. *La reforma agraria en el desarrollo económico de México.* Mexico: Instituto Mexicano de Investigaciones Económicas, 1969.
Almada, Francisco. *Diccionario de historia, geografía y biografía Chihuahuenses.* Chihuahua, Mexico: Talleres Gráficos del Gobierno del Estado, 1927.
———. *La rebelión de Tomochí.* Chihuahua, Sociedad Chihuahuense de Estudios Históricos, 1938.
———. *La revolución en el estado de Chihuahua.* 2 vols. Chihuahua: Biblioteca del Inst. Nacional de Estudios Históricos de la Revolución Mexicana, 1964.
Bailey, David C. "Revisionism and the Recent Historiography of the Mexican Revolution." *Hispanic American Historical Review* 58 (February 1978): 62–79.
Bazant, Jan. *Historia de la deuda exterior de México (1823–1946).* Mexico: El Colegio de México, 1968.
Beals, Carlton. *Porfirio Díaz: Dictator of Mexico.* Philadelphia: J. B. Lippincott Co., 1932.
Bellingeri, Marco, and Gil Sánchez, Isabel. "Las estructuras agrarias bajo el Porfiriato." In *México en el siglo XIX (1821–1910): Historia económica y de la estructura social,* edited by Ciro Cardoso, pp. 315–38. Mexico: Editorial Nueva Imagen, 1980.

Benjamin, Thomas. "Revolución interrumpida—Chiapas y el interinato presiden-cial—1911." *Historia mexicana* 30 (July–September 1980): 79–88.

Benjamin, Thomas, and Ocasio-Meléndez, Marcial. "Organizing the Memory of Modern Mexico: Porfirian Historiography in Perspective, 1880s–1980s." *Hispanic American Historical Review* 64 (May 1984): 323–64.

Berry, Charles R. *The Reform in Oaxaca, 1856–76: A Microhistory of the Liberal Revolution.* Lincoln: University of Nebraska Press, 1981.

Brading, D. A., ed. *Caudillo and Peasant in the Mexican Revolution.* Cambridge: Cambridge University Press, 1980.

Buve, Raymond Th. J. "Peasant Movements, Caudillos and Landreform [*sic*] during the Revolution (1910–1917) in Tlaxcala, Mexico." *Boletín de estudios latinoamericanos y del Caribe* 18 (June 1975): 112–52.

Camp, Roderic A. *Entrepreneurs and Politics in Twentieth-Century Mexico.* New York: Oxford University Press, 1989.

Chevalier, Francois. *L'Amerique Latine de l'indépendance a nos jours.* Paris: Presses Universitaires de France, 1977.

Coatsworth, John H. "Del atraso al subdesarrollo: La economía mexicana de 1800 al 1910." Chicago, 1984.

———. "Los orígenes del autoritarismo moderno en México." *Foro internacional* 16 (October–December 1975): 205–32.

Coerver, Don M. "The Perils of Progress: The Mexican Department of Fomento during the Boom Years, 1880–1884." *Inter-American Economic Affairs* 31 (Autumn 1977): 41–62.

———. *The Porfirian Interregnum: The Presidency of Manuel González of Mexico, 1880–1884.* Fort Worth: Texas Christian University Press, 1979.

Contreras, Mario, and Tamayo, Jesús, eds. *México en el siglo XX, 1900–1913: Textos y documentos.* Mexico: Universidad Nacional Autónoma de Mexico, 1975.

———. *México en el siglo XX, 1913–1920: Textos y documentos.* Mexico: Universidad Nacional Autónoma de México, 1976.

Córdova, Arnaldo. *La formación del poder político en México.* Mexico: Serie Popular Era, 1972.

Cosío Villegas, Daniel, gen. ed. *Historia moderna de México.* 9 vols. Mexico: Editorial Hermes, 1955–72.

Cossío, José L. *¿Como y por quienes se ha monopolizado la propiedad rústica en México?* Mexico: Tip. Mercantial, 1911.

Covarrubias, José. *Varios informes sobre tierras y colonización.* Mexico: Imp. y Foto. de la Sec. de Fomento, 1912.

Creelman, James. "Díaz Speaks." In *The Age of Porfirio Díaz: Selected Readings,* edited by Carlos B. Gil, pp. 78–81. Albuquerque: University of New Mexico Press, 1977.

Dale-Lloyd, Jane. *El proceso de modernización capitalista en el noroeste de Chihuahua (1880–1910).* Mexico: Universidad Iberoamericana, Departamento de Historia, 1987.

De la Maza, Francisco F. *Código de colonización y terrenos baldíos de la república mexicana.* Mexico: Ofc. Tip. de la Sec. de Fomento, 1893.

De la Peña, Sergio. *La formación del capitalismo en México*. Mexico: Siglo Veintiuno Editores, 1975.

Diario Oficial. Mexico City.

Díaz, Luis Miguel, ed. *México y las comisiones internacionales de reclamaciones*. 2 vols. Mexico: Universidad Nacional Autónoma de Mexico, 1983.

Díaz-Polanco, Héctor, and Laurent Guye Montandon. *Agricultura y sociedad en el Bajío (S. XIX)*. Mexico: Centro de Investigación para la Integración Social, 1984.

Díaz Rugama, Adolfo. *Prontuario de leyes, reglamentos, circulares y demás disposiciones vigentes relativas a . . . la Secretaría de Fomento. . . .* Mexico: Eduardo Dublan, 1895.

Eckstein, Susan. *The Impact of Revolution: A Comparative Analysis of Mexico and Bolivia*. Beverly Hills: Sage Publications, 1976.

Falcón, Romana. "¿Los orígenes populares de la revolución de 1910? El caso de San Luis Potosí." *Historia mexicana* 29 (October–December 1979): 197–240.

Fix Zamudio, Graciela. "Joaquín D. Casasús: Humanista mexicano del siglo XIX." Licenciado thesis, Universidad Nacional Autónoma de México, 1963.

Florescano, Enrique. "Ensayo de Interpretación." In *Latin America: A Guide to Economic History, 1830–1930,* edited by Roberto Cortes Conde and Stanley Stein, pp. 435–55. Berkeley: University of California Press, 1977.

———. "La influencia del estado en la historiografía mexicana." *i Siempre!,* Aug. 18, 1976, pp. iv–xi.

———. *Orígen y desarrollo de los problemas agrarios de México, 1500–1821*. Mexico: Ediciones Era, 1976.

Friedrich, Paul. *Agrarian Revolt in a Mexican Village*. Chicago: University of Chicago Press, 1977.

Gibson, Charles. *The Aztecs under Spanish Rule: A History of the Indians of the Valley of Mexico, 1519–1810*. Stanford: Stanford University Press, 1964.

Gilly, Adolfo. "La guerra de clases en la revolución mexicana (Revolución permanente y auto-organización de las masas)." In *Interpretaciones de la revolución mexicana,* pp. 21–54. Mexico: Editorial Nueva Imagen, 1980.

González Calzada, Manuel. *Historia de la revolución mexicana en Tabasco*. Mexico: Consejo Editorial del Gobierno del Estado de Tabasco, 1981.

González Navarro, Moisés. *La colonización en México, 1877–1910*. Mexico: Talleres de Imp. de Estampillas y Valores, 1960.

Guerra, François-Xavier. *México: Del antiguo régimen a la revolución*. 2 vols. Mexico: Fondo de Cultura Económica, 1988. Trans. of *Le Mexique: De l'ancien régime à la révolution*. Paris: L'Harmattan, 1985.

Hall, Linda B. *Alvaro Obregón: Power and Revolution in Mexico, 1911–1920*. College Station: Texas A&M University Press, 1981.

Hamilton, Nora. *The Limits of State Autonomy: Post-Revolutionary Mexico*. Princeton: Princeton University Press, 1982.

Hansen, Roger D. *The Politics of Mexican Development*. Baltimore: Johns Hopkins University Press, 1971.

Hernández Chávez, Alicia. "La defensa de los finqueros en Chiapas—

1914–1920." *Historia mexicana* 28 (January–March 1979): 335–69.

Hobsbawn, E. J. *The Age of Capital, 1848–1875.* New York: New American Library, 1975.

Los hombres del centenario. Mexico: n.p., 1910.

Hu-Dehart, Evelyn. "Pacification of the Yaquis in the Late Porfiriato: Development and Implications." *Hispanic American Historical Review* 54 (February 1974): 337–59.

Jacobs, Ian. *Ranchero Revolt: The Mexican Revolution in Guerrero.* Austin: University of Texas Press, 1982.

Joseph, G. M. *Revolution from Without: Yucatán, Mexico and the United States, 1880–1924.* Cambridge: Cambridge University Press, 1982.

Katz, Friedrich. "Mexico: Restored Republic and Porfiriato, 1867–1910." In vol. 5 of *The Cambridge Modern History of Latin America.* Edited by Leslie Bethell. Cambridge: Cambridge University Press, 1984–.

———. *The Secret War in Mexico.* Chicago: University of Chicago Press, 1981.

Kitching, Gavin. *Development and Underdevelopment in Historical Perspective: Populism, Nationalism and Industrialization.* London: Routledge, 1989.

Knight, Alan. *The Mexican Revolution.* 2 vols. Cambridge: Cambridge University Press, 1986.

———. "Peasant and Caudillo in Revolutionary Mexico, 1910–1917." In *Caudillo and Peasant in the Mexican Revolution.* Edited by D. A. Brading. Cambridge: Cambridge University Press, 1980.

Kozhevar, E. *Informe sobre la república mexicana presentado al Consejo de Tenedores de Bonos Extranjeros.* Translated by Joaquín D. Casasús. Mexico: Ofc. Tip. de la Sec. de Fomento, 1887.

Leal, Juan Felipe. "El estado y el bloque en el poder en México: 1867–1914." *Historia mexicana* 23 (April–June 1974): 700–721.

Limantour, José Ives. *Apuntes sobre mi vida pública.* Mexico: Porrúa, 1965.

López Rosado, Diego G. *Historia y pensamiento económico de México.* 2 vols. Mexico: UNAM, Instituto de Investigaciones Económicos, 1968.

Loyola Díaz, Rafael. *La crisis Obregón-Calles y el estado mexicano.* Mexico: Siglo Veintiuno Editores, 1980.

Márquez Montiel, Joaquín. *Hombres célebres de Chihuahua.* Mexico: Editorial Jus, 1953.

McBride, George McC. *Land Systems of Mexico.* New York: American Geographical Society, 1923.

Mejía Fernández, Miguel. *Política agraria en México en el siglo XIX.* Mexico: Siglo Veintiuno Editores, 1979.

Menegus Bornemann, Margarita. "Ocoyoacac—Una comunidad agraria en el siglo XIX." *Historia mexicana* 30 (July–September 1980): 33–78.

Mestre Ghigliazza, Manuel. *Apuntes para una relación cronológica de los gobernantes de Tabasco.* Mérida: n.p., 1934.

Mexico. Ministerio de Fomento. *Anuario estadístico de la república mexicana, 1893.* Mexico: Ofc. Tip. de la Sec. de Fomento, 1894.

———. Ministerio de Fomento. *Anuario estadístico de la república mexicana, 1897.* Mexico: Ofc. Tip. de la Sec. de Fomento, 1898.

———. Secretariá de Fomento. *Anuario estadístico de la república mexicana, 1901.* Mexico: Ofc. Tip. de la Sec. de Fomento, 1902.

———. Secretariá de Fomento. *Anuario estadístico de la república mexicana, 1906.* Mexico: Ofc. Tip. de la Sec. de Fomento, 1910.

———. Secretaría de Fomento, Colonización, Industria y Comercio. *Boletín semestral de la estadística de la república mexicana,* no. 3 (1889).

———. Secretaría de Fomento, Colonización e Industria. Dirección General de Estadística. *Boletin,* no. 1 (1912).

———. Secretaría de Fomento, Colonización, Industria y Comercio. *Informe que rinde el Secretario de Fomento a la honorable Cámara de Diputados sobre colonización y terrenos baldíos.* Mexico: Ofc. Tip. de la Sec. de Fomento, 1885.

———. Secretaría de Fomento, Colonización, Industria y Comercio. *Memoria de la Secretaría de Estado y del Despacho de Fomento, Colonización, Industria y Comercio de la República Mexicana escrita por el ministro del ramo, C. Manuel Siliceo, para dar cuenta con ella al Congreso Constitucional.* Mexico: Imp. de Vicente García Torres, 1857.

———. Secretaría de Fomento, Colonización, Industria y Comercio. *Memoria presentada al congreso de la unión . . . corresponde a los años trascurridos de diciembre de 1877 a diciembre de 1882.* Mexico: Ofc. Tip. de la Sec. de Fomento, 1885.

———. Secretaría de Fomento, Colonización, Industria y Comercio. *Memoria presentada al congreso de la union . . . 1883–85.* Mexico: Ofc. Tip. de la Sec. de Fomento, 1887.

———. Secretaría de Fomento, Colonización e Industria. *Memoria presentada al congreso de la unión . . . 1892–96.* Mexico: Ofc. Tip. de la Sec. de Fomento, 1897.

———. Secretaría de Fomento, Colonización e Industria. *Memoria presentada al congreso de la unión . . . 1897–1900.* Mexico: Imp. y Fototipia de la Sec. de Fomento, 1908.

———. Secretaría de Fomento, Colonización e Industria. *Memoria presentada al congreso de la unión . . . 1901–4.* Mexico: Imp. y Fototipia de la Sec. de Fomento, 1909.

———. Secretaría de Fomento, Coloniación e Industria. *Memoria presentada al congreso de la unión . . . 1905–7.* Mexico: Imp. y Fototipia de la Sec. de Fomento, 1909.

———. Secretaría de Fomento, Colonización e Industria. *Memoria presentada al congreso de la unión . . . 1907–8.* Mexico: Imp. y Fototipia de la Sec. de Fomento, 1910.

———. Secretaría de Fomento, Colonización e Industria. *Memoria presentada al congreso de la unión . . . 1908–9.* Mexico: Imp. y Fototipia de la Sec. de Fomento, 1910.

———. Secretaría de Fomento, Colonización e Industria. *Memoria presentada al congreso de la unión . . . 1909–10.* Mexico: Imp. y Fototipia de la Sec. de Fomento, 1910.

———. Secretaría de Fomento, Colonización e Industria. *Memoria presentada al congreso de la unión . . . 1910–11.* Mexico: Imp. y Fototipia de la Sec. de Fomento, 1912.

————. Secretario de Programación y Presupuesto. Instituto Nacional de Estadística Geografía e Informática. *Agenda estadística 1984.*

Meyer, Jean-A. "Le Mexique a la veille de la révolution de 1910: Crédibilité des statistiques agraires," *Revue historique* 562 (April–June 1987).

Meyer, Lorenzo. "Historical Roots of the Authoritarian State in Mexico." In *Authoritarianism in Mexico,* edited by José Luis Reyna and Richard S. Weinert, pp. 3–22. Philadelphia: Institute for the Study of Human Issues, 1977.

Molina Enríquez, Andrés. *Los grandes problemas nacionales.* Mexico: Imp. de A. Carranza e Hijos, 1909. Reprint. Mexico: Centro de Estudios Históricos del Agrarismo en Mexico, 1984.

Moreno García, Heriberto, coord. *Despúes de los latifundios (la desintegración de la gran propiedad agraria en México)* N.p.: Colegio de Michoacán, 1982.

Nell, Edward. "Economics: The Revival of Political Economy." In *Ideology in Social Science: Readings in Critical Social Theory,* edited by Robin Blackburn, pp. 76–95. New York: Pantheon Books, 1972.

North, Douglass C. *Structure and Change in Economic History.* New York: W. W. Norton & Co., 1981.

North, Douglass C., and Thomas, Robert Paul. *The Rise of the West: A New Economic History.* Cambridge: Cambridge University Press, 1973.

Orozco, Wistano Luis. *Legislación y jurisprudencia sobre terrenos baldíos.* Mexico: Imp. de El Tiempo, 1895.

Paz, Ireneo. *Los hombres prominentes de México.* Mexico: La Patria, 1888.

————. *México actual: Galería de contemporaneos.* Mexico: Ofc. Tip. de la Patria, 1898.

Phipps, Helen. *Some Aspects of the Agrarian Question in Mexico: A Historical Study.* Austin: University of Texas, 1925.

Romero, Matias. *Mexico and the United States.* New York: G. P. Putnam's Sons, 1898.

Rosenzweig, Fernando. "El desarrollo económico de México de 1877 a 1911." *Trimestre económico* 32 (July/September 1965): 405–54.

————. "El proceso político y el desarrollo económico de México." *Trimestre económico* 29 (1962): 513–30.

San Juan Victoria, Carlos, and Velázquez Ramírez, Salvador. "El estado y las políticas económicas en el Porfiriato." In *México en el siglo XIX (1821–1910): Historia económica y de la estructura social,* edited by Ciro Cardoso, pp. 277–312. Mexico: Editorial Nueva Imagen, 1980.

Schryer, Frans J. "Peasants and the Law: A History of Land Tenure and Conflict in the Huasteca." *Journal of Latin American Studies* 18 (November 1986) 2:283–311.

————. *The Rancheros of Pisaflores: The History of a Peasant Bourgeoisie in Twentieth-Century Mexico.* Toronto: University of Toronto Press, 1980.

Sierra, Justo. *The Political Evolution of the Mexican People.* Translated by Charles Ramsdell. Austin: University of Texas Press, 1969.

Silva Herzog, Jesús. *El agrarismo mexicano y la reforma agraria.* Mexico: Fondo de Cultura Económica, 1959.

Simpson, Eyler N. *The Ejido: Mexico's Way Out*. Chapel Hill: University of North Carolina Press, 1937.

Smith, Robert Freeman. *The United States and Revolutionary Nationalism in Mexico, 1916–1932*. Chicago: University of Chicago Press, 1972.

Solís, Leopoldo. *La realidad económica mexicana: Retrovisión y perspectivas*. Rev. ed. Mexico: Siglo Veintiuno Editores, 1981.

Southworth, John R. *El estado de Sinaloa, Mexico*. San Francisco: Press of the Hicks-Judd Co., 1898.

Stevens, Donald Fithian. "Agrarian Policy and Instability in Porfirian Mexico." *Americas* 34 (October 1982): 153–66.

Tannenbaum, Frank. *Peace by Revolution: An Interpretation of Mexico*. New York: Columbia University Press, 1933. Reprint. *Peace by Revolution: Mexico after 1910*. New York: Columbia Paperback Edition, 1966.

Taylor, William B. *Drinking, Homicide and Rebellion in Colonial Mexican Villages*. Stanford: Stanford University Press, 1979.

Teuffer, Salvador. *El Departamento Agrario en la integración de Tabasco a la revolución mexicana*. Mexico: n.p., 1935.

Thompson, E. P. *Whigs and Hunters: The Origin of the Black Act*. New York: Pantheon Books, 1975.

Tilly, Charles. *From Mobilization to Revolution*. Reading, Mass.: Addison-Wesley, 1978.

Tutino, John. *From Insurrection to Revolution in Mexico: Social Bases of Agrarian Violence, 1750–1940*. Princeton: Princeton University Press, 1986.

United States–Mexican Commission. *Proceedings of the United States–Mexican Commission Convened in Mexico City, May 14, 1923*. Washington: U.S. Government Printing Office, 1925.

Valadés, José C. *El Porfirismo: Historia de un régimen*. Vol. 2, *El crecimiento: I*. Mexico: José Porrúa e Hijos, 1941–48. Reprint of vols. 2 and 3. Mexico: Universidad Nacional Autónoma de México, 1977.

Velasco, Alfonso Luis. *Geografía y estadística de la república mexicana*. 20 vols. Mexico: Ofc. Tip. de la Sec. de Fomento, 1889–1898.

Villamar, Aniceto. *Las leyes federales vigentes sobre tierras, bosques, aguas, ejidos, colonización y el gran registro de propiedad*. 2d ed. Mexico: Herrero Hermanos, 1910.

Walker, David W. *Kinship, Business and Politics: The Martínez del Río Family in Mexico, 1824–1867*. Austin: University of Texas Press, 1986.

———. "Porfirian Labor Politics: Working Class Organizations in Mexico City and Porfirio Díaz, 1876–1902." *Americas* 37 (January 1981): 257–90.

Wasserman, Mark. *Capitalists, Caciques, and Revolution: The Native Elite and Foreign Enterprise in Chihuahua, Mexico, 1854–1911*. Chapel Hill: University of North Carolina Press, 1984.

Wells, Allen. "Family Elites in a Boom-and-Bust Economy: The Molinas and Peóns of Porfirian Yucatán." *Hispanic American Historical Review* 62 (May 1982): 224–53.

Whetten, Nathan L. *Rural Mexico*. Chicago: University of Chicago Press, 1948.

Wodon de Sorinne, Guillermo. *La colonización de México*. Mexico: Sec. de Fomento, 1902.

Wolf, Eric. *Europe and the People without History.* Berkeley: University of California Press, 1982.

Womack, John, Jr. *Zapata and the Mexican Revolution*. New York: Vintage Books, 1970.

Index

Lightning Source UK Ltd.
Milton Keynes UK
UKHW040746290919
350631UK00003B/216/P